Lecture Notes Mathematics

A collection of informal reports and seminars
Edited by A. Dold, Heidelberg and B. Eckmann, Zürich

319

Conference on Group Theory

University of Wisconsin-Parkside 1972

Edited by R. W. Gatterdam and K. W. Weston
University of Wisconsin-Parkside, Kenosha, WI/USA

Springer-Verlag
Berlin · Heidelberg · New York 1973

AMS Subject Classifications (1970): 20-02, 20B25, 20C05, 20C30, 20D10, 20E10, 20E15, 20E20, 20E35, 20F05, 20F10, 20F15, 20F25, 20F45, 20F99, 20J05, 20J10, 20K15, 20K20, 20K30, 20K99, 02F35, 02F47, 20D15, 20E25, 20F50, 20E30, 20G20, 20H20

ISBN 3-540-06205-X Springer-Verlag Berlin · Heidelberg · New York
ISBN 0-387-06205-X Springer-Verlag New York · Heidelberg · Berlin

Offsetdruck: Julius Beltz, Hemsbach/Bergstr.

ACKNOWLEDGEMENTS

These proceedings are the outgrowth of a conference on group theory sponsored by the University of Wisconsin-Parkside. The conference was held at Wingspread Conference Center in Racine, Wisconsin, on June 28 to 30, 1972 and chaired by the editors.

The editors wish to thank the University of Wisconsin-Parkside and in particular Vice Chancellor Otto F. Bauer for the necessary resources; Mrs. Kay Mauer and the Wingspread staff for the hospitality; the contributors and participants for the success.

R.W. Gatterdam
K.W. Weston

TABLE OF CONTENTS

ON THE STRUCTURE OF HECKE ALGEBRAS

Nelo D. ALLAN

University of Wisconsin-Parkside

Our purpose is to study the problem of writting down their structure equation for a large family of such algebras and, at some extent, try to fit our results inside of a possible general theory.

1. Preliminaries. Let G be a submonoid of a group $\hat{G}$, i.e., G is multiplicatively closed and has the same unity of $\hat{G}$, and let H be a subgroup of $\hat{G}$ which is contained in G. We can form the free Z-module, $R=R(G,H)$, Z rational integers, generated by the double cosets modulo H of the elements of G; we are interested in the case where R has an structure of an associative algebra and this happens in the case where for every element g of G the double coset $\bar{g}=HgH$ is the union of finitely many right cosets Hg_i, because in this case we can define product as follows: If $g,g'\in G$, we set $\bar{g}\cdot\bar{g}'=\Sigma\, m_i\bar{h}_i$ where $HgHg'H=\cup\, Hh_iH$, disjoint union, with m_i being the number of pairs (j,m) such that $Hg_jg'_m=Hh_i$, where $Hg'H=\cup Hg'_m$. This happens in the case where G is contained in the commensurability group, $\tilde{G}(H)$, of H, i.e., the subgroup of $\hat{G}$ consisting of all elements g such that $g^{-1}Hg$ and H are commensurable, i.e., $g^{-1}Hg\cap H$ has finite index in both $g^{-1}Hg$ and H. In this case we call $R=R(G,H)$ the Hecke algebra of (G,H), and if K is a

(*) The author apologizes for presenting a different topic from his lecture.

field we set $R(G,H;K) = R(G,H) \otimes_Z K$. In several cases it is possible to put in G a structure of locally compact unimodular group and take H as an open compact subgroup; here R can be interpreted as the convolution algebra of all continuous, Z-valued, two-sided H-invariant functions on G. There are two interesting situations. The first one is when G is a finite group; here $H=\{e\}$ yields the group algebra $Z[G]$ of G. The second is the case where k is the quotient field of a Dedekind domain Λ, with all the residue class fields finite, G is a reductive group linear algebraic and defined over k. G being linear, is a matrix group, hence we can consider the subgroup G_k (resp. G_Λ) consisting of all matrices g having its entries in k (resp. g and g^{-1} have entries in Λ) which in several cases is contained in $\tilde{G}(G_\Lambda)$ and take H as anything commensurable to G_Λ ; for instance this happens in the case where k is either a p-adic field or is a real field with the group of real points of G having no compact component. Let p be a prime ideal in Λ and $k(p)$ be the p-adic completion of k, $\Lambda(p)$ being the ring of integers of $k(p)$ is also the p-adic completion of Λ. We shall assume that G has the weak approxiamtion property: the p-adic closure $\overline{G}_k$ of G_k is $G_{k(p)}$ for all finite primes p in Λ. If G_Λ generates an order in the ring of all matrices over k, then G_Λ is open compact. For any H commensurable to G_Λ we can regard the algebra $R(G_{k(p)},\overline{H})$ as a subalgebra of $R(G_k,H)$; hence $R(G_k,H)$ contains a subalgebra isomorphic to the restricted tensor product of all such $R(G_{k(p)},\overline{H})$, for all finite primes p; it is known that the equality of those two algebras occur in the case of a simply connected Chevalley group, Λ is a principal ideal domain and H is the stabilizer of an admissible lattice, [4] . In view of the above considerations one is lead to study the local case: G is the set of k-rational points of a reductive linear group defined over k, discrete valuation locally compact field, and H open-compact subgroup of G . In order to cover the classical groups we sometimes allow k to be non commutative, making the proper adjustment.

2. Quick survey of known results

Hecke operators were introduced by Hecke in his studies of Zeta functions and Eisenstein series associated to the modular group; shortly afterwards Peterson generalized these studies to principal congruence subgroups of the modular group. In 1959, Shimura (c.f.[11]) considered the algebra generated by the Hecke operators and called it a Hecke algebra. Tamagawa (c.f.[13]) in order to study the local factors of the Zeta function of a division algebra, determined the structure of $R(G_k, G_\Lambda; \mathbb{C})$, $\mathbb{C}$=complex numbers, $G=Gl_n(k)$, with k local division algebra with maximal order Λ. In this case R is a polynomial algebra. Satake (c.f.[10]) generalized this result to the classical reductive groups in their usual representation and also gave conditions on (G,H) in order to have in the general case R commutative. Satake uses Fourier analysis in a reductive group. Next Goldman and Iwahori extended these results to simply connected Chevalley groups, and later Iwahori (c.f. [3]) calculated the algebra of (G,B) , B= Borel subgroup of the Chevalley group G over a finite field; his calcula tion are based in the BN-pair structure of G and R ~k[W] (Tits), W = group of the BN-pair, K algebraically closed with characteristic not dividing the order of G . Iwahori-Matsumoto (c.f.[5]) extended these calculations to the generalized Tits system in a Chevalley group; they proved that R(G,B) is a twisted tensor algebra for B a parahoric subgroup of G , i.e.,

(*) For more details we refer to Iwahori's Boulder lectures (c.f.[4])

B is the pull-back in G_Λ of a Borel subgroup of $G_\Lambda/H(1)$, $H(1)$ being the first principal congruence subgroup of G_Λ, regarding this quotient as an algebraic group over the residue class field. More recently MacDonald (c.f. [7]) using anlytic methods calculated the structure of R for a p-adic Chevalley group, with H being the stabilizer of an admissible lattice; R is a commutative ring. In the case of a finite group it is well known, (c.f.[3]), that R(G,H;K) is semisimple, provided that the characteristic of K does not divide the order of G; it is trivial to show that this algebra is a Frobenius algebra if the characteristic does not divide the index [G:H]. In the case of $G=Gl_2$ the work of Sally-Shalika [9], and Silberger [12], have as consequence that R(G,H;K) is a direct sum of finite dimensional matrix alge bras. There is a conjecture that this result holds for any p-adic group G, (c.f. [12], p. 16). This is the one of the many steps needed to generalize Langland's theory to those group.

3. Algebras of subgroups

In our recent note (c.f. [1]) we observed that if there exists in G an open compact subgroup U, a k-closed subgroup T consisting only of semisimple elements, N^+ and N^- maximal k-closed unipotent subgroups normalized by T, satisfying the conditions H-1 and H-2 written below for a normal subgroup H of U, then the algebra R is finitely generated; more precisely, it is generated by Z[U/H] and Z[D]. Our conditions are:

(H-1) There exists a semigroup D in T such that G=UDU, and for all d in D, $dU^+d^{-1} \subset U^+$ and $d^{-1}U^-d \subset U^-$, where $U^+ = N^+ \cap U$ and $U^- = N^- \cap U$.

Let H be a normal subgroup of U; we set $U_o^+ = U^+ \cap H$ and $U_o^- = U^- \cap H$, and also $V = T \cap H$ and assume that D normalizes V.

(H-2) $H = U_o^+ V U_o^-$, $TN^+ \cap H = VU_o^+$, and for all d in D , we have $dU_o^+ d^{-1} \subset U_o^+$ and $d^{-1} U_o^- d \subset U_o^-$.

If moreover both $d^{-1}Hd$ and dHd^{-1} are contained in U for all the generators d of D , then we can write recurrence formulas for the products which do not follow from group relations, namely the products of the type $\overline{d_1} \cdot \overline{gd_2}$, $d_1, d_2 \in D$, $g \in U$; if moreover $U = U^+ W U^+$ it suffices to list these products for $g \in WU^+$. This is very simple to do in the case of Gl_2 for the principal congruence subgroups. Our conditions (H-1),(H-2) are satisfied for principal congruence subgroups of the following classical groups: linear group, symplectic group, unitary groups, and non-split orthogonal groups. Although we have not yet had the opportunity of putting together the results available in the literature, it seems that they are also sati fied for simply connected Chevalley groups with $H = E_q$ (c.f. [8]). An immediate consequence of our calculations is that any representation of $R(G,H)$ such that the image of D are units , is necessarily abelian, provid ed that k is commutative and U is generated by U^+ and U^- . These conditions have others consequences: the groups for which they are satisfied for the congruence subgroups $H(s) = \{ g \in U \mid g \equiv 1 \text{ modulo } p^s \}$, and having the congruence subgroup property are such that all $R(G,H)$ are aces sible by brute force calculation. We shall discuss this now. Let H and H' be any two open-compact subgroups of G such that H' is normal in H ; we set $m = [H:H']$, $H = \cup H'g_i$. We can regard the H-invariant functions on G as being H'-invariant and we get an immersion of $R(G,H)$ into $R(G,H')$ under the mapping $m\psi$ where $\psi(g)$ is the characteristic function of HgH in $R(G,H')$. Let us set $\bar{\bar{g}} = H'gH'$ and $\underline{e}(H) = \Sigma \bar{\bar{g}}_i$. Then $\psi(\bar{g})$

$= (n^2/m(g))\underline{e}(H).\bar{\bar{g}}.\underline{e}(H)$, with $m(g) = \ell(g)r(g)\lambda(g)$, $\ell(g)$ (resp. $r(g)$) being the order of the left stabilizer (resp. right stabilizer) of $\bar{\bar{g}}$ in $\bar{\bar{H}}$ (we denote it by $L(g)$ (resp. $R(g)$)), and $\lambda(g)$ is the number of distinct pairs (i,j) , where $g_i, g_j \in H$ and $\bar{\bar{g}}_i \cdot \bar{\bar{g}} = \bar{\bar{g}} \cdot \bar{\bar{g}}_j$, with $\bar{\bar{g}}_i$ taken modulo $L(g)$ and $\bar{\bar{g}}_j$ taken modulo $R(g)$. Now given $g,h \in G$, we set $[g,h;i] = \Sigma a_j m(t_j)\bar{t}_j$, if $\bar{\bar{g}} \cdot \bar{\bar{g}}_i \cdot \bar{\bar{h}} = \Sigma a_j \bar{\bar{t}}_j$, then we have

$$(1) \qquad \bar{g} \cdot \bar{h} = m(g)^{-1} m(h)^{-1} \sum_{i=1}^{m} [g,h;i]$$

If G is finite, this formula does not say much more than [3]; it become interesting in our situation. For instance in the particular case where $G^* = BDB = U^+DU^-(1)$, $B = H(1)U^+$, $U^-(1) = U^- \cap H(1)$, and $H(s) \subset H \subset H(r)$, $2r \geq s$, then we have the following formula for all the products in $R(G^*,H)$

$$(2) \qquad \bar{g} \cdot \bar{h} = \frac{m(gh)}{m(g)m(h)} \sum_{i=1}^{m} \bar{\alpha}_i \cdot \overline{gh} \cdot \bar{\beta}_i$$

with $\alpha_i, \beta_i \in H(r)$ and $gg_ih = \alpha_i gh\beta_i$. Hence we have a handy formula to calculate the products in the algebra generated by D; more simplification can occur in particular cases. If we add an extra condition on $\{G,U,D,H(s)\}$ which is very natural and essencially says that for any d in D every element $u \in U^+$ (resp. U^-) can be written as a product $u = u_1 u_2$, (resp. $u_2 u_1$) with $\bar{u}_1 \in R(d)$ (resp. $L(d)$) and the other commutes with d , with some others tecnical assumptions, then there exists N such that

$$m(td_1^{a_1} \ldots d_n^{a_n} t') = m(td_1^{b_1} \ldots d_n^{b_n} t')$$

where $t,t' \in U$, $d_1,\ldots,d_n$ are all generators of D and $a_i = b_i$ if either

$a_i \leq N$ or d_i is central, and $b_i = N$ otherwise. Let us look now to the case of Gl_2 ; here we set $d =$ diagonal $(\pi, 1)$, π being the generator of p , and we also set $g_i = u_i f_i n_i$, $u_i \in U_o^+$, $f_i \in V$ and $n_i \in U_o^-$. Then

$$(3) \qquad \overline{d^a} \cdot \overline{d^b} = bb' \underline{g}(F') \cdot \overline{d^{a+b}} \qquad , \quad a,b \geqslant s$$

with $F = \{f_1, \ldots, f_\nu\}$ are the distinct f_j ordered if necessary in such way that $F' = \{f_1, \ldots, f_\mu\}$ are all the f_j's such that $\{\overline{f_j d^{a+b}}\}$ are all the distinct terms in (2) , b is the number of indices i such that $\overline{f}_i = \overline{1}$ and $b' = [F:F']$. The other products are not difficult to list and shall be published elsewhere. Also in a forthcoming note we shall apply these results to discuss finite generation of R.

BIBLIOGRAPHY

1- N. Allan, Hecke Rings of congruence subgroups, Bull. Amer. Math. Soc. vol. 74,#4(1972) 541-545.

2- Harish-Chandra and G. v. Dijk, Harmonic Analysis on Reductive p-adic groups, Lecture Notes in Math., 162(1970), Springer Verlag, New York.

3- N. Iwahori, On the Structure of a Hecke ring of a Chevalley group over a finite field, J. Fac. of Sci. Univ. Tokyo , vol. 10(1964) 215-236.

4- N. Iwahori, Generalized Tits System on p-adic Semisimple Groups, Proc. Symp. in Pure Math., vol. IX,(1966), 71-83, Amer. Math. Soc. Providence , R.I.

5- N. Iwahori and M. Matsumoto, On some Bruhat decomposition and the structure of the Hecke rings of the p-adic Chevalley groups, Publ. Math. I.H.E.S., # 25(1965), 5-48.

6- R. Langlands and H. Jacquet , Automorphic forms on Gl(2), Lecture Notes in Math. vol. # 114,(1970), Springer Verlag, New York.

7- I. G. MacDonald, Spherical functions on a p-adic Chevalley group, Bull. Amer. Math. Soc. , vol. 70(1968) 520-525.

8- M. Matsumoto, Sur les sous-groupes arithmetiques des groupes semisimple deployes, Ann. Scient. Ec. Norm. Sup. 4° serie, t.2, (1969)1-62.

9- P. Sally and J. Shalika, Characters of the discrete series of representations of Sl(2) over a local field, Proc. Nat. Acad. Sc. U.S.A. vol 61 (1968), 1231-1237.

10- I. Satake, Theory of spherical functions on reductive algebraic groups over p-adic fields, Plub. Math. I.H.E.S.,#18 (1964),5-69.

11- G. Shimura, Sur les integrales attachees aux formes automorphes, J. Math. Soc. of Japan,vol 11 (1959), 291-311.

12- A. Silberger, PGl_2 over the p-adics, Lecture Notes in Math., vol. 166 (1970), Springer Verlag, New York.

13- T. Tamagawa, On the ζ-function of a division algebra, Ann. of Math. vol 77 (1963), 387-405.

On Locally Extended Residually Finite Groups

R.B.J.T. Allenby and R. J. Gregorac

Introduction

Several recent papers have investigated the group theoretical properties of residual finiteness and certain of its generalizations in connection with various group 'product' constructions. A few natural questions remain and it is the object of this note to give an example which settles some of them. The generalizations of residual finiteness that we look at are given in

Definition 1. (i) A group G is said to be (locally) extended residually finite if to each (finitely generated) subgroup H of G and to each element $g \in G \setminus H$ there exists a homomorphism ϕ of G onto a finite group in which $g\phi \notin H\phi$.

(ii) A group G is said to be π_c if a similar conclusion holds whenever H is a cyclic subgroup of G.

Both definitions in (i) can be found in [1] while (ii) is obviously equivalent to that given in [2].

We denote the class of (residually) finite groups by (R)F, the class of (locally) extended residually finite groups by (L)ERF, and the class of π_c groups again by π_c. Then we clearly have

$$F \subseteq ERF \subseteq LERF \subseteq \pi_c \subseteq RF.$$

Reading from the left the first and second inequalities are strict; for the first take the infinite cycle and for the second a two generator free group using the theorem of M. Hall [3] that free groups are LERF. That the fourth inequality is strict is asserted in [4] though no example is given. An easy example is given by the group $A = \{a,b; b^{-1}ab = a^2\}$. This group is a finitely generated metabelian group, hence in RF by a well known theorem of P. Hall. It is trivially not in π_c since $a \notin \{a^2\}$ and yet if $A\phi$ is any finite group of order n we have $a\phi = b^{-n}\phi \cdot a\phi \cdot b^n\phi = a^{2n}\phi \in \{a^2\phi\}$. This leaves the third inequality. Our main result shows it too is strict.

In [5], [2] and [1] (see also [6]) it has been shown that RF, π_c, LERF are closed with respect to the taking of the free product of two groups. Trivially F is not so closed; nor is ERF, where the free group on two generators again suffices. With respect to direct products (of two groups) F is trivially closed while the facts that ERF, π_c and RF are closed are in [7], [2] and [5]. This leaves the problem for LERF. We summarize this briefly as follows

$A,B \in \mathfrak{X}$	F	ERF	LERF	π_c	RF
$A*B \in \mathfrak{X}$	no	no	yes	yes	yes
$A \times B \in \mathfrak{X}$	yes	yes	?	yes	yes

In fact we show that ? can be replaced by no.

Finally a result on generalized free products. Evans and Boler [8] have recently proved the

<u>Theorem 2.</u> Let $A \underset{U}{*} B$ be the generalized free product of A,B amalgamating H where A,B $\in$ RF and H is a retract of A and of B. Then $A \underset{H}{*} B \in$ RF. (Here H is a <u>retract</u> of G means H has a normal complement in G).

Theorem 2 may be readily extended as follows.

<u>Theorem 2'.</u> Suppose A,B $\in \pi_c$ and H is a retract of both A and B. Then $A \underset{H}{*} B \in \pi_c$.

Furthermore, if "π_c" is replaced throughout Theorem 2' by "LERF", our example will show the resulting statement is not true.

To prove Theorem 2' one only needs to apply the methods of Stebe's paper [2] and consider several cases; in each case one maps $A \underset{H}{*} B$ onto factor groups of the form $A/N \underset{H'}{*} B/M$, $H' = HN/N \cong HM/M$, which are in π_c. The only problem is to verify these factor groups are themselves in π_c. To this end we note

<u>Lemma 3.</u> If A and B are LERF (π_c) and U is finite, then the generalized free product $A \underset{U}{*} B$ is LERF (π_c). (c.f. Theorem 3 [10]).

<u>Proof:</u> Since groups which are LERF (π_c) are residually finite are normal subgroups X and Y of finite index in A and B respectively such that $U \cap X = 1 = U \cap Y$. Thus there exists a canonical homomorphism ϕ from $A \underset{U}{*} B$ onto a finite group P

generated by A/X and B/Y such that ker $\varphi \cap A = X$ and ker $\varphi \cap B = Y$. It follows from the subgroup theorem for generalized free products that ker φ is an ordinary free product of conjugates of subgroups of X and Y with a free group (see Lemma 1 [11]). Thus ker φ is a free product of LERF (π_c) groups and hence LERF (π_c). Therefore $A \underset{U}{*} B$ is a finite extension of a LERF (π_c) group, and thus, as may be easily shown, must itself be a LERF (π_c) group, proving the lemma.

Finally, it should be noted that Theorem 2' can be easily extended to a generalized free product G of arbitrarily many groups amalgamating a single subgroup which is a retract of each factor of G and, also, Theorem 2 of [8] on knot groups immediately extends to π_c groups.

The Example

Let $F = \{a,b: \ \}$, $G = \{x,y: \ \}$ be 2-generator free groups. The elements of $D = F \times G$ will be written as ordered pairs. We show $D \notin$ LERF. For let H be the subgroup of D generated by the elements (a,x), (b,y), (1,r) where $r = y^{-1}x^2yx^{-3}$. Clearly any conjugate of (1,r) by $(1,g) \in G$ is in H and so

$$H \cap G \supseteq \{(1,r)\}^G,$$

where X^G means the normal closure of X in G. Conversely if $(1,s) \in H \cap G$ then (1,s) is a product of powers of the generators of H. Since the product of the powers of a and b occurring in the first coordinates must be trivial, if all the occurrences of (1,r) in this product are deleted the product of the remaining terms are also the identity. Thus

$$H \cap G = \{(1,r)\}^G.$$

Now $G/(H \cap G)$ is a finitely generated non-hopf group [9] and as such cannot be residually finite. Thus there is an element $t \notin H \cap G$ such that t belongs to every normal subgroup N of G which has finite index in G and contains $H \cap G$. Hence $t \notin H$, but if M is a normal subgroup of finite index in $F \times G$ then $t \in (M \cap G)(H \cap G)$. Thus $t \notin H$, but $t\sigma \in H\sigma$ for all homomorphisms σ of $F \times G$ onto a finite group.

This example thus shows that LERF is not closed under direct products. Also, since π_c is closed under direct products, $F \times G$ is in π_c but not in LERF.

Finally we note that if C_1 and C_2 are infinite cycles and if we set $A = C_1 \times F$, $B = G \times C_2$ and $T = A \underset{F=G}{*} B$, then $T = (C_1 * C_2) \times F$ is not a LERF group. Further A and B both have F(=G) as a retract. Thus we will have a counterexample to the possible extension of Theorem 2 to LERF groups if only we can prove $C_1 \times F$ is LERF. More than this is proved by the following extension of a result of Mal'cev's [7] used in [8].

<u>Theorem 4</u>. Let $G = NM$, $N \cap M = 1$, be a splitting extension of the normal finitely generated subgroup N by the group M. If

N is ERF and M is ERF (LERF), then G is ERF (LERF). If N and M are both π_c groups, then G will be a π_c group.

<u>Proof</u>: Following the proof of Mal'cev's result in [8] (Lemma 1) it is easy to see that G is either a finite extension of an ERF (LERF) group or a finite extension of a π_c group, respectively, proving the theorem. Hence $F \times C_1$ is LERF since F is LERF and C_1 is ERF. It is quite interesting to compare the result of Theorem 4 with our example.

<u>Further Observations</u>

(i) It is now easy to show that the only verbal product with respect to which LERF is closed is the free product. By definition the verbal product A(V)B corresponding to the variety $\underline{\underline{V}}$ is the group $A*B/V(A*B) \cap [A,B]$.
Letting $\bar{A},\bar{B}$ denote the images of A,B in this factor group we see that $[V(\bar{A}),\bar{B}]$ is the trivial subgroup of A(V)B. If A,B are both free groups of rank 2 and if $\underline{\underline{V}}$ is not the variety of all groups then in A(V)B, $V(\bar{A})$ is a non-trivial free group of infinite rank and $V(\bar{A})$ and $\bar{B}$ generate their direct product $V(\bar{A}) \times \bar{B}$. This is not a LERF group-hence neither is A(V)B.

(ii) In [10] B. Baumslag defined the concept of 'fully residually free' (f.r.f.) He showed that if G is a residually free group without center then either G is f.r.f. or else has a subgroup of the form $F \times G$ where F,G are both free groups of rank 2. It thus appears that if G is a centerless residually free group the presence of the LERF condition is enough to force G to be f.r.f. whereas the presence of the π_c condition is not.

(iii) Finally we mention two extensions of results in [14] and [12].

<u>Theorem 5.</u> If the groups A and B are polycyclic by finite, then $A \underset{W}{*} B$ is LERF, whenever there exists a subgroup U of finite index in W such that U is normal in both A and B. (c.f. Theorem 9 [14]).

<u>Proof</u>: Since polycyclic groups are LERF [7], A and B are LERF.

Let H be a finitely generated subgroup of $G = A \underset{W}{*} B$ and $g \in G \backslash H$. If $g \notin HU$, then $gU \notin HU/U$ in the group $A/U \underset{W/U}{*} B/U$ which is LERF by Lemma 3, so $g\theta \notin H\theta$ in some finite factor group $G\theta$ of G. So suppose $g = hu$, $h \in H$, $u \in U \backslash H$. Let N be a subgroup of finite index in U containing $H \cap U$ but not u. Let C be the intersection of the (finitely many) normal subgroups S of U such that $|U : S| = |U : N|$. Then C is a characteristic subgroup of finite index in U, with $u \notin C(H \cap U)$. Let ϕ be the canonical homomorphism of G onto the LERF group

$$G/C \cong A/C \underset{W/C}{*} B/C.$$

Since $u\phi \notin H\phi$, there must again be a homomorphism θ of G onto a finite group such that $u\theta \notin H\theta$, completing the proof.

Corollary 5.1. If A and B are both extensions of finite groups by finitely generated abelian groups, then any generalized free product $A \underset{W}{*} B$ of A and B is LERF. (c.f. Theorem 6 [12]).

Proof: Here A and B are also polycyclic by finite ([13], Theorem 2.1) and by the proof of Theorem 6 [12], W^r is normal in both A and B for some integer r, so Theorem 5 applies to complete the proof.

Theorem 6 (See [12]). If A and B are polycyclic by finite groups and U is a cycle, then $G = A \underset{U}{*} B$ is π_c.

The proof of this again follows using Stebe's methods [2], Lemma 3, and the following lemma found in [12].

Lemma 7. Let $A \underset{U}{*} B$ be a generalized free product of polycyclic by finite groups, with $U = gp\{c\}$ an infinite cycle. Then there exist normal subgroups N_A, N_B, of A, B respectively such that

$$N_A \cap U = N_B \cap U = gp\{c^{mt}\}$$

where t is any given integer and m is an integer determined by $A \underset{U}{*} B$.

References

1. R. G. Burns, "On finitely generated subgroups of free products", J. Austral. Math. Soc., 12(1971), 358-364.

2. P. Stebe, "Residual finiteness of a class of knot groups", Comm. Pure and Applied Math., 21(1968), 563-583.

3. M. Hall, "Coset representation in free groups", Trans. Amer. Math. Soc., 67(1949), 421-432.

4. P. Stebe, "On free products of isomorphic free groups with a single finitely generated amalgamated subgroup", J. Algebra, 11(1969), 359-362.

5. K. W. Gruenberg, "Residual properties of infinite soluble groups", Proc. London Math. Soc. Ser. 3, 7(1957), 29-62.

6. N. S. Romanovskii, "On the residual finiteness of free products with respect to subgroups", Izv. Akad. Nauk SSSR Ser. Mat., 33(1969), 1324-1329 (Russian).

7. A. I. Mal'cev, "On homomorphisms onto finite groups", Ivanov. Gos. Ped. Inst. Učen. Zap., 18(1958), 49-60 (Russian).

8. B. Evans and J. Boler, "The free products of residually finite groups amalgamated along retracts is residually finite", submitted to Proc. Amer. Math. Soc.

9. G. Baumslag and D. Solitar, "Some two-generator one-relator non-hopfian groups", Bull. Amer. Math. Soc., 68(1962), 199-201.

10. B. Baumslag, "Residually free groups", Proc. London Math. Soc. (3), 17(1967), 402-18.

11. R. J. Gregorac, "On generalized free products of finite extensions of free groups", J. London Math. Soc., 41(1966), 662-666.

12. J. L. Dyer, "On the residual finiteness of generalized free products", Trans. Amer. Math. Soc., 133(1968), 131-143.

13. P. F. Smith, "Quotient rings of group rings", J. of the London Math. Soc., 3(1971), 645-660.

14. G. Baumslag, "On the residual finiteness of generalized free products of nilpotent groups", Trans. Amer. Math. Soc., 106(1963), 193-209.

On the conjugacy problem for knot groups

K. I. Appel

Abstract

A method is developed to show word and conjugacy problems solvable for a large class of knot groups. The class includes groups of a large number of non-alternating knots for which no previous conjugacy results have been obtained. The method involves a modification of the small cancellation diagrams of Lyndon and Schupp applied to Wirtinger presentations of knot groups. The crucial tool is a dual to each small cancellation diagram consisting of a set of curves in the plane of a projection of the knot.

It is hoped that this approach will enable one to show that all knot groups have solvable conjugacy problem.

EXCEPTIONAL PRIMES IN VARIETIES

S. Bachmuth

I.

The exceptional role which various primes assume in certain varieties is at the moment very obscure. As an introduction, let me list some interesting examples which come immediately to mind.

(a) $p = 2$ in the variety $\mathfrak{M}_3$ of 3-metabelian groups.

Indeed, if a group in $\mathfrak{M}_3$ has no elements of order 2, then it is metabelian [1]; and if we allow such elements, then the group need not be metabelian [2].

(b) $p = 5$ in the variety $\mathcal{E}_3$ of third Engel groups.

If a group in $\mathcal{E}_3$ has no elements of order 5, then using results of Heinekin [3] and Gupta and Weston [4], Gupta [5] has shown that such a group is solvable. But without such a restriction, an element of $\mathcal{E}_3$ need not be solvable [6].

(c) $p = 2$ in any locally finite variety which contains an infinite simple group.

If $\mathfrak{V}$ is such a variety, then a theorem of Kovács [7] states that $\mathfrak{V}$ must contain an infinite number of finite simple groups. Since all non-abelian finite simple groups have even order [8], we see that $p = 2$ must occur in $\mathfrak{V}$.

Of course (c) is really a question as to whether a locally finite variety can contain an infinite simple group. We know [9], that there exists a locally finite subvariety of the variety $\mathcal{B}_5$ (defined by the law $x^5 = 1$) and which contains a group G which is perfect (equal to its commutator subgroup). Thus there are many examples of groups which are locally finite, perfect, and having bounded exponent in which $p = 2$ occurs. (Interesting examples of such perfect groups satisfying the law $x^n = 1$, where $n = 2^\alpha 5n'$, and α and n' are positive integers, are the wreath products of G by A, where A is any non-abelian finite simple group.) Thus, it may be reasonable at this time to ask whether there are also simple groups with these same properties. Because of Kovács' Theorem, the existence of such a simple group would have several interesting consequences contrary to long-standing conjectures.

(i) There exists locally finite varieties containing an infinite number of finite simple groups, and

(ii) For every natural number k, there exists a finite simple group requiring at least k generators.

II. Exceptional primes in $\mathcal{E}_4$.

The outstanding question concerning the Engel varieties $\mathcal{E}_n$ (defined by the law $[x, y, y, \ldots, y] = 1$, where y is repeated n times) is whether for $n \geq 4$, $\mathcal{E}_n$ are locally solvable. Heinekin [3] has shown that $\mathcal{E}_3$ is locally solvable, but for $n > 3$ the question is open. However, for $n \geq 3$, we now know that $\mathcal{E}_n$ is a non-solvable variety and so to start right out and ask even if $\mathcal{E}_4$ is locally solvable is unreasonable. What we should do is try to add further conditions which will force solvability on $\mathcal{E}_4$. In other words, ask what primes are exceptional in $\mathcal{E}_4$.

There has been important work by Higgins [10] and refinements in Walkup [11] on analogous problems for Lie rings. In fact, M. Putcha [12] has shown that a Lie ring of characteristic 7 satisfying the fourth Engel condition is nilpotent. This result combined with Lemma 3.1 on page 92 of Walkup [11] improves Theorem 4 of Higgins [10] to the conclusion that if a Lie ring L satisfies the fourth Engel condition and has characteristic prime to 5!, then L is nilpotent.

This is all we have to go by in searching for reasonable conjectures for $\mathcal{E}_4$. Although we do not know the precise relationships between Lie rings and groups, we can nonetheless guide ourselves to some degree. It is my conjecture that 2 and 5 are the exceptional primes in $\mathcal{E}_4$.

If we are eventually to develop a very clear overall picture, it is important for us to know whether or not 7 is exceptional in $\mathcal{E}_4$. There is every reason to believe that 7 is exceptional for $\mathcal{E}_5$, the precise conjecture being that a group (Lie ring) of exponent $p \geq 5$ (characteristic p) satisfying the (p-2) Engel condition need not be solvable. The case $p = 5$ was established in [6] for groups and in [9] for Lie rings.

An unexpected consequence of showing that 2 is not an exceptional prime in $\mathcal{E}_4$ would be that the variety $\mathcal{B}_4$ (of groups having exponent 4) is a solvable variety. (This is a consequence of reductions made in [4].) It is for this reason that I conjecture that 2 is exceptional in $\mathcal{E}_4$, because the evidence seems to point in the direction of the non-solvability of $\mathcal{B}_4$. A proof of this may be close at hand and I shall discuss this in a later section.

Finally, let me mention a problem which is perhaps more accessible than the fourth Engel one. It's solution may even throw some light on $\mathcal{E}_4$.

If we denote $\mathcal{M}_i$ as the variety whose i generator groups are metabelian, then clearly

$$\mathcal{M}_2 \supset \mathcal{M}_3 \supset \mathcal{M}_4 = \mathcal{M}_5 = \cdots \ .$$

(I. D. Macdonald [1] has shown that $\mathfrak{M}_3$ is defined by the law $[[x,y],[x,z]] = 1$, and G. Higman [13] has shown that $\mathfrak{M}_2$ is defined by the law $[[x,y],[x^{-1},y]] = 1$.) B. H. Neumann [2] established that the above containments are in fact proper. While establishing that $\mathfrak{M}_3$ may also be defined by the Jacobi identity $[x,y,z][z,x,y][y,z,x] = 1$, Bachmuth and Lewin [14] established that the free group in $\mathfrak{M}_3$ is a central extension of a direct product of cyclic groups of order 2 by a free metabelian group. Hence $\mathfrak{M}_3$ is very close to being metabelian. But in dramatic contrast, $\mathfrak{M}_2$ is a non-solvable variety [6]. The question therefore is what are the exceptional primes of $\mathfrak{M}_2$? i.e., minimal conditions that will force solvability on $\mathfrak{M}_2$?

In the next section I will recall an example which illustrates the startling surprises that can happen even in metabelian varieties.

III. A very exceptional example.

This short section is really a diversion from our main line of thought, but I take this opportunity to mention it for several reasons. First, because of the exceptional and startling nature of the result, it perhaps should find a place in this talk. But more important, this result is not as well known as it deserves to be and I welcome the opportunity to publicize it. I am hopeful that by bringing this example to the attention of the many experts assembled here, it will be scrutinized in the near future and eventually explained to our satisfaction.

Let $M = F/F''$ be a free metabelian group on three generators (any number larger than two would do), and M_n the nth term of the lower central series beginning $M_1 = M$. In my thesis I calculated those automorphisms of M/M_n which are induced by M, and in so doing I was able to show that if an automorphism of M/M_4 is induced by an automorphism of M (not all are), then it can be induced by an automorphism of the absolutely free group F. This is also true for $n = 3$, but is a simple result in this case, and of course the natural question which came to mind but which I couldn't answer was if the same was true for all n?

The answer given by Chein [15] was yes, <u>except</u> when $n = 5$!!!

IV. Groups of exponent four.

Showing that $\mathcal{B}_4$ is a non-solvable variety appears tantalizingly ripe at present because of the formal similarity with the problem of $\mathcal{K}_5$, the Kostrikin variety of exponent 5. Virtually a mere change in characteristic from 5 to 2. It's true we can no longer use the same ideal H of [9]. This would be tantamount to imposing the third Engel condition when going over to the group, and it is known [4] that this forces solvability on a group of exponent 4. However, we need only factor by the ideal $\mathfrak{J}$ defined in Section 4 of [9]. In characteristic 5, our associative ring R is non-solvable modulo $\mathfrak{J}$, and so all we need do is show the same is true when we change the characteristic to 2. One may now try to show non-solvability in a somewhat similar fashion as is done for characteristic 5, or else may one perhaps be able to deduce non-solvability using the hypothesis of non-solvability in characteristic 5? This latter approach of trying to avoid the real problem was for a time considered by H. Mochizuki and myself. Although it didn't work for us, I still feel there may be merit to this approach.

Returning to the more conventional procedures, characteristic 2 causes many difficulties. For example, we cannot get some information by partial linearizations which we were able to get in characteristic 5, etc. But some recent work of Mochizuki, Weston, Yuan, and myself has led to a startling development.

One of the family of polynomials which generate the ideal $\mathfrak{I}$ is the symmetric polynomial $S(x,y,z)$ which is the sum of all possible monomials in x, y, z of total degree three and degree one in each of the variables. Of course, if we take for x, y, z all equal elements, then because our characteristic is 2 we cannot conclude that $x^3 = 0$ in R as would be the case in characteristic 5. Suppose we now enlarge our ideal $\mathfrak{I}$ to include the cubes of all indeterminates in R. We presumably no longer have a Lie substitution ideal, but what now emerges may be a parallel phenomena as in [9]. The homogeneous subspaces of our ideal, consisting of exactly degree three in any indeterminate which appears, all have dimension one! So once again there appears to be certain subspaces for which no new information is generated by the substitution of Lie elements (where allowable, i.e., in $\mathfrak{I}$) in place of indeterminates. Once again it may be possible to push through a proof of non-solvability without knowing all or even most of the relations determined by our ideal. At present we have a uniform procedure for going to higher dimensions from lower dimensions, but we have only ad hoc methods for assembling elements within the same subspace. Until and unless we find a uniform procedure (algorithm) for dealing with these elements, we will be stymied in a conclusive proof. More exciting than the actual proof, however, will be the possibility of discovering a common theme which will enable us to predict untried situations.

References

[1] MacDonald, I. D.: On certain varieties of groups. Math. Z. 76, 270-282 (1961); II. Math Z. 78, 175-188 (1962).

[2] Neumann, B. H.: On a conjecture of Hanna Neumann. Proc. Glasgow Math. Assoc. 3, 13-17 (1956).

[3] Heinekin, H.: Engelsche Elemente der Länge drei. Ill. J. Math. 5, 681-707 (1961).

[4] Gupta, N., Weston, K.: On groups of exponent four. J. Algebra 17, 59-66 (1971).

[5] Gupta, N.: Third Engel 2-groups are solvable. Canad. Math. Bull. (to appear).

[6] Bachmuth, S., Mochizuki, H.: Third Engel groups and the MacDonald-Neumann conjecture. Bull. Austral. Math. Soc. 5, 379-386 (1971).

[7] Kovács, L. G.: Varieties and finite groups. J. Austral. Math. Soc. 10, 5-19 (1969).

[8] Feit, W., Thompson, J G.: Solvability of groups of odd order. Pacific J. Math. 13, 775-1029 (1963).

[9] Bachmuth, S., Mochizuki, H., Walkup, D.: Construction of a nonsolvable group of exponent 5, Word Problems (edited by W W. Boone, R. C. Lyndon, F. B. Cannonito. North-Holland, Amsterdam; in press).

[10] Higgins, P. J.: Lie rings satisfying the Engel condition. Proc. Camb. Philos. Soc. 50, 381-390 (1956).

[11] Walkup, D. W.: Thesis, University of Wisconsin, 1963.

[12] Putcha, M. S.: On Lie rings satisfying the fourth Engel condition. Proc. Amer. Math. Soc. 28, 355-357 (1971).

[13] Higman, G.: Some remarks on varieties of groups. Quart. J. Math. Oxford 10, 165-178 (1959).

[14] Bachmuth, S., Lewin, J.: The Jacobi identity in groups. Math. Z. 83, 170-176 (1964).

[15] Chein, O.: IA automorphisms of free and free metabelian groups. Comm. Pure and Applied Math. 21, 605-629 (1968).

Group with Free Subgroups of Finite Index

Daniel E. Cohen

Serre conjectured, and Stallings and Swan proved [6], [7], that a torsion-free group with a free subgroup of finite index is free. Solitar [5] made a conjecture about arbitrary groups with a free subgroup of finite index, and proved the conjecture for finitely generated groups. This paper proves the conjecture for countable groups, and gives some information on the uncountable case.

Let Φ denote the class of free groups, and $\mathfrak{F}$ the class of finite groups. Then $\Phi\mathfrak{F}$ denotes the class of groups with a free normal subgroup of finite index, and this is easily seen to coincide with the class of groups having a free subgroup of finite index. If $G \in \Phi\mathfrak{F}$ then G is finitely generated iff it has a finitely generated free subgroup of finite index (using Schreier's formula).

Solitar's conjecture. *If $G \in \Phi\mathfrak{F}$ then G is an HNN group whose base is a tree product of finite groups, and whose associated subgroups each lie in some vertex group of the tree product.*

Plainly if $H \lhd G$, H free, G/H finite, then any finite subgroup of G is isomorphic to a subgroup of G/H. Conversely Solitar shows that if G is an HNN group whose base group is a tree product of finite groups *of bounded order* and whose associated subgroups each lie in some vertex group of the tree product, then $G \in \Phi\mathfrak{F}$.

In terms of Serre's theory of graphs of groups [4], Solitar's conjecture may be restated as:

If $G \in \Phi\mathfrak{F}$, then G is the fundamental group of a graph of groups for which all vertex groups are finite.

By Serre's main theorem in [4], this can again be restated as:

If $G \in \Phi\mathfrak{F}$, then G acts on a tree in such a way that the stabiliser of each vertex is finite.

The best way to prove the conjecture would be to determine the relevant tree, but I am unable to do this. However a result of Serre (Theorem C in [1]) immediately tells us that if $G \in \Phi\mathfrak{F}$ $cd_Q\, G = 1$ (where Q denotes the rationals). Much of the theory of Stallings and Swan can now be applied.

1. Graphs of Groups

Definition. A graph of groups $(\mathcal{G}, X)$ consists of

(i) a connected graph X (two different edges may have the same vertices; also the two vertices of an edge may coincide).

(ii) a group G_v for each vertex v and a group G_e for each edge e;

(iii) a monomorphism $G_e \to G_v$ for each edge e of X and each vertex v of e (if both vertices of e are v, we require two monomorphisms $G_e \to G_v$).

If Y is a connected subgraph of X, we have the induced graph of groups $(\mathcal{G}|Y, Y)$.

Let T be a maximal tree in X. Then the fundamental group of $(\mathcal{G}, X)$ relative to T, written $\pi(\mathcal{G}, X, T)$, is defined to be the HNN group whose base is the tree product over T of the groups G_v with the two images of G_e amalgamated for each edge e of T, where the associated pairs of subgroups are the two images of G_e for each edge e not in T. Serre [4] shows that, up to isomorphism, $\pi(\mathcal{G}, X, T)$ is independent of T, so we shall write $\pi(\mathcal{G}, X)$ where convenient.

It is easy to see that if Y is a connected subgraph of X such that $Y \cap T$ is a maximal tree of Y then there is a natural monomorphism $\pi(\mathcal{G}|Y, Y, Y\cap T) \to \pi(\mathcal{G}, X, T)$, and we shall identify $\pi(\mathcal{G}|Y, Y, Y \cap T)$ with its image in $\pi(\mathcal{G}, X, T)$. In particular, each vertex group G_v will be regarded as a subgroup of $\pi(\mathcal{G}, X, T)$.

We shall call the graphs of groups $(\mathcal{G}, X)$ and $(\mathcal{H}, X)$ conjugate if

(i) for each vertex v and edge e of X we have $H_v = G_v$ and $H_e = G_e$;

(ii) for each edge e and vertex v of e, the monomorphism $H_e \to H_v$ is the monomorphism $G_e \to G_v$ followed by conjugation by an element of G_v (if the vertices of e coincide, the two monomorphisms $H_e \to H_v$ may arise from the monomorphisms $G_e \to G_v$ by conjugation by two different elements).

Lemma 1. Let $(\mathcal{H}, X)$ and $(\mathcal{G}, X)$ be conjugate. For any maximal tree T there is an isomorphism $\theta : \pi(\mathcal{G}, X, T) \to \pi(\mathcal{H}, X, T)$ with the property that for each vertex w of X there is an element c_w of $\pi(\mathcal{H}, X, T)$ such that $\theta x = c_w^{-1} x c_w$ for $x \in G_w$; further given a vertex v, we may choose θ so that $c_v = 1$.

Note that we are identifying G_w with its images in $\pi(\mathcal{G}, X, T)$ and $\pi(\mathcal{H}, X, T)$.

Proof Let T_n denote the subtree of T consisting of all vertices of T whose distance in T from v is at most n, together with the corresponding edges. Let $K_n = \pi(\mathcal{G}|T_n, T_n, T_n)$, $K = \pi(\mathcal{G}|T, T, T)$ and $L_n = \pi(\mathcal{H}|T_n, T_n, T_n)$, $L = \pi(\mathcal{H}|T, T, T)$. Then $K = \bigcup K_n$, $L = \bigcup L_n$, and we shall define $\theta_{n+1} : K_{n+1} \to L_{n+1}$ extending $\theta_n : K_n \to L_n$. As $T_0 = v$, we take θ_0 to be the identity. Assume θ_n defined.

Each edge e of $T_{n+1} - T_n$ has one vertex v_{1e} in T_n and another vertex v_{2e} in $T_{n+1} - T_n$. Let $\alpha_{ie} : G_e \to G_{v_{ie}}$ and $\beta_{ie} : G_e \to G_{v_{ie}}$, $i = 1, 2$ be the monomorphisms in $\mathcal{G}$ and $\mathcal{H}$.

By hypothesis, there exist elements a_{ie} in $G_{v_{ie}}$ such that

$$\beta_{ie} x = a_{ie}^{-1}(\alpha_{ie} x) a_{ie}, \qquad \text{for} \qquad x \in G_e.$$

Now K_{n+1}, as a tree product, is obtained from the free product of K_n and the groups $G_{v_{2e}}$, $e \in T_{n+1} - T_n$, by adding the relations $\alpha_{1e}x = \alpha_{2e}x$ for $x \in G_e$, $e \in T_{n+1} - T_n$, and similarly for L_{n+1}. Hence we may define θ_{n+1} by $\theta_{n+1} = \theta_n$ on K_n and $\theta_{n+1}x = c_w^{-1}xc_w$ for $w = v_{2e}$ provided we have $\theta_n(\alpha_{1e}x) = c_w^{-1}(\alpha_{2e}x)c_w$ in L_{n+1} for $x \in G_e$.

By our inductive hypothesis, $\theta_n(\alpha_{1e}x) = c_{v_{1e}}^{-1}(\alpha_{1e}x)c_{v_{1e}}$ in L_n and by definition of H we have $a_{1e}^{-1}(\alpha_{1e}x)a_{1e} = a_{2e}^{-1}(\alpha_{2e}x)a_{2e}$ in L_{n+1}. Consequently we need only take $c_w = a_{2e}\, a_{1e}^{-1}\, c_{v_{1e}}$ for $w = v_{2e}$, to obtain a homomorphism $\theta_{n+1} : K_{n+1} \to L_{n+1}$ extending θ_n. Similarly we obtain a homomorphism $\phi_{n+1} : L_{n+1} \to K_{n+1}$ extending $\phi_n : L_n \to K_n$. Easy calculations show ϕ_{n+1} and θ_{n+1} are inverse isomorphisms if ϕ_n and θ_n are.

The isomorphisms θ_n give an isomorphism $\theta : K \to L$. Now for each edge e of $X - T$ let α_{ie}, β_{ie}, $i = 1, 2$ be the monomorphisms of G_e into the corresponding vertex groups for G and H, and let a_{ie} be as before. Then $\pi(\mathcal{G}, X, T)$ is the HNN group

$$<K,\ x_e;\ x_e^{-1}(\alpha_{1e}g)x_e \ = \ \alpha_{2e}g,\ g \in G_e>$$

where e runs over the edges of $X - T$, and similarly $\pi(\mathcal{H}, X, T)$ is the HNN group $<L,\ y_e;\ y_e^{-1}(\beta_{1e}g)y_e = \alpha_{2e}y,\ y \in G_e>$. Hence we can extend $\theta : K \to L$ to a homomorphism $\pi(\mathcal{G}, X, T) \to \pi(\mathcal{H}, X, T)$ mapping x_e onto u_e where $u_e^{-1}(\theta\alpha_{1e}g)u_e = \theta(\alpha_{2e}g)$ for all $g \in G_e$ (and all $e \in X - T$). Hence we should take $u_e = c_{v_{1e}}^{-1}\, a_{1e}\, y_e\, a_{2e}^{-1}\, c_{v_{2e}}$ to obtain the map $\theta : \pi(\mathcal{G}, X, T) \to \pi(\mathcal{H}, X, T)$. Similarly we obtain $\phi : \pi(\mathcal{H}, X, T) \to \pi(\mathcal{G}, X, T)$, and check that ϕ is the inverse of θ.

Let X be a connected graph, X_i a collection of mutually disjoint connected subgraphs of X. We may define a connected graph Y whose edges are the edges of $X - \bigcup X_i$, whose vertices are the vertices of $X - \bigcup X_i$ together with one vertex v_i for each index i, and such that if e is an edge of $X - \bigcup X_i$ then the vertices of e in Y are v and w if $v, w \notin X_i$, are v and v_i if $v \notin X_i, w \in X_i$, and are v_i and v_j if $v \in X_i, w \in X_j$. We call Y the graph obtained from X by contracting the X_i, and will also say that X is a graph obtained from Y by expanding the X_i.

If T_i is a maximal tree in X_i for all i, and T is a maximal tree in X such that $T \cap X_i = T_i$ for all i (such trees T exist) then the graph S obtained from T by contracting the T_i is a maximal tree of Y.

With the above notation, let $(\mathcal{G}, X)$ be a graph of groups. We shall define a graph of groups $(\mathcal{H}, Y)$ which we will say is obtained from $(\mathcal{G}, X, T)$ by contracting the X_i (and will also say that $(\mathcal{G}, X)$ is obtained from $(\mathcal{H}, Y)$ by expanding the X_i). For each edge e of Y let $H_e = G_e$. For each vertex v of Y not of the form v_i let $H_v = G_v$, and let $H_{v_i} = \pi(\mathcal{G}|X_i, X_i, T_i)$. If the edge e in Y has vertex v not of the form v_i the monomorphism $H_e \to H_v$ is to be the monomorphism $G_e \to G_v$, while if v_i is a vertex of e in Y there will be a vertex x of e lying in X_i, and $H_e \to H_{v_i}$ is to be $G_e \to G_x \to \pi(\mathcal{G}|X_i, X_i, T_i)$.

It is easy to see that there is an isomorphism between $\pi(\mathcal{G}, X, T)$ and $\pi(\mathcal{H}, Y, S)$ identifying the subgroup $\pi(\mathcal{G}|X_i, X_i, T_i)$ of $\pi(\mathcal{G}, X, T)$ with the subgroup H_{v_i} of $\pi(\mathcal{H}, Y, S)$.

Lemma 2. Let $(\mathcal{G}, X)$ be a graph of groups such that for each vertex v of X there is a graph of groups (G_v, X_v) and maximal tree T_v of X with $G_v = \pi(G_v, X_v, T_v)$. Suppose each edge e of X and vertex v of e there is a vertex x of X such that the image of G_e in G_v lies in a conjugate

<u>of the subgroup G_x of $\pi(\mathcal{G}_v, X_v, T_v)$. Then there is a graph of groups which is obtained from a conjugate of $\pi(\mathcal{G}, X)$ by expanding the X_v.</u>

Remarks 1. By an (easy) result of Serre the hypothesis on the groups G_e holds if each G_e is finite.

2. X_v may consist of a single point.

<u>Proof</u> Replacing $(\mathcal{G}, X)$ by a conjugate graph of groups we may assume that to each edge e of X with vertex v we have chosen a vertex x of X_v such that the image of G_e in G_v lies in G_x.

We define a graph Y whose vertices are the vertices of $\bigcup X_v$ and whose edges are the edges of $X \cup \bigcup X_v$. If an edge e in X_v has vertices x, y, in X_v, then the edge e in Y is to have x, y as vertices. If e in X has vertices v, w, by hypothesis we have chosen vertices x of X_v, y of X_w such that the images of G_e in G_v and G_w lie in G_x and G_y. We then require x and y to be the vertices of e in Y.

It is now clear how to find a graph of groups $(\mathcal{H}, Y)$ such that $(\mathcal{G}, X)$ is obtained from $(\mathcal{H}, Y)$ by contracting the X_v.

If T is a maximal tree in X, then the edges of $T \cup \bigcup T_v$ and the corresponding vertices form a maximal tree in Y.

<u>Definition.</u> Let H be a subgroup of a group G. We say H is a <u>vertex</u> of G, written $H \vee G$, if there is a graph of groups $(\mathcal{G}, X)$ with maximal tree T and vertex v such that:

(i) G_w is finite for each vertex $w \neq v$;

(ii) G_e is finite for each edge e;

(iii) there is an isomorphism $\pi(\mathcal{G}, X, T) \to G$ sending G_v onto H.

Note that $1 v G$ iff $\exists$ a graph of groups $(\mathcal{G}, X)$ with each vertex group finite such that G is isomorphic to $\pi(\mathcal{G}, X)$. For if such $(\mathcal{G}, X)$ exists we may show $1 v G$ using the graph $(\mathcal{G}_1, X_1)$ where X_1 is X together with an extra vertex and an extra edge joining this vertex to an edge of X, the groups corresponding to this extra vertex and edge being trivial. Hence Solitar's conjecture becomes: <u>if $G \in \Phi\mathcal{F}$, then $1 v G$</u>. If G is torsion-free then $H v G$ iff $G = H*F$ where F is free. So if G is torsion-free then $1 v G$ iff G is free. Note also that if G is finite, then $G = H *_H G$ shows that $1 v G$.

<u>Lemma 3. Let G_α be a collection of groups indexed by the ordinal numbers $\leq \lambda$, such that $G_\alpha \, v \, G_{\alpha+1}$ for all $\alpha < \lambda$ and $G_\beta = \bigcup_{\alpha<\beta} G_\alpha$ for all limit ordinals $\beta \leq \lambda$. Then $G_0 \, v \, G_\lambda$.</u>

Corollary. If $H \leq K \leq G$ and $H \, v \, K$, $K \, v \, G$, then $H \, v \, G$.

<u>Proof</u>. By hypothesis for each $\alpha < \lambda$ we have a graph of groups $(\mathcal{G}_\alpha, X_\alpha)$ with maximal tree T_α and vertex v_α such that there is an isomorphism $\pi(\mathcal{G}_\alpha, X_\alpha, T_\alpha) \to G_{\alpha+1}$ sending G_{v_α} to G_α, and with all edge-groups and all vertex-groups except G_{v_α} finite.

Suppose we have a graph of groups $(\mathcal{H}_\alpha, Y_\alpha)$ with maximal tree S_α such that there is an isomorphism $\theta_\alpha : \pi(\mathcal{H}_\alpha, Y_\alpha, S_\alpha) \to G_\alpha$. By lemmas 1 and 2 there will be a graph of groups $(\mathcal{H}_{\alpha+1}, Y_{\alpha+1})$ and maximal tree $S_{\alpha+1}$ such that $Y_\alpha \subseteq Y_{\alpha+1}$, $S_\alpha = Y_\alpha \cap S_{\alpha+1}$, $\mathcal{H}_\alpha = \mathcal{H}_{\alpha+1} \mid Y_{\alpha+1}$ such that there is an isomorphism

$$\theta_{\alpha+1} : \pi(\mathcal{H}_{\alpha+1}, Y_{\alpha+1}, S_{\alpha+1}) \to G_{\alpha+1}$$

extending θ_α and such that the groups corresponding to a vertex or edge of $Y_{\alpha+1} - Y_\alpha$ will be finite.

We now define $\mathcal{H}_\alpha$, Y_α, S_α and an isomorphism $\theta_\alpha : \pi(\mathcal{H}_\alpha, Y_\alpha, S_\alpha) \to G_\alpha$ for all $\alpha \leq \lambda$ by transfinite induction. Start with Y_0 a point. Suppose the definitions made for all $\alpha < \rho$ with the property that for all $\alpha < \beta < \rho$ we have $Y_\alpha \subseteq Y_\beta$, $S_\alpha = Y_\alpha \cap S_\beta$, $\mathcal{H}_\alpha = \mathcal{H}_\beta \mid Y_\alpha$, and $\theta_\alpha = \theta_\beta \mid \pi(\mathcal{H}_\alpha, Y_\alpha, S_\alpha)$. If $\rho = \alpha+1$ the last paragraph gives the definition of $\mathcal{H}_\rho$, Y_ρ, S_ρ, θ_ρ. If ρ is a limit ordinal define $Y_\rho = \bigcup_{\alpha<\rho} Y_\alpha$, $S_\rho = \bigcup_{\alpha<\rho} S_\alpha$, with the obvious definition of $\mathcal{H}_\rho$. Since $G_\rho = \bigcup_{\alpha<\rho} G_\alpha$ and $\pi(\mathcal{H}_\rho, Y_\rho, S_\rho) = \bigcup_{\alpha<\rho} \pi(\mathcal{H}_\alpha, Y_\alpha, S_\alpha)$, we can define θ_ρ.

In particular θ_λ is defined and we have $G_0 \vee G_\lambda$ since the groups corresponding to all vertices and edges of $Y_\lambda - Y_0 = \bigcup_{\alpha<\lambda} (Y_{\alpha+1} - Y_\alpha)$ will be finite.

<u>Lemma 4. Let $G = G_1 *_K G_2$, where K is finite. If $1 \vee G_2$, then $G_1 \vee G$.</u>

<u>Proof</u>. $G = \pi(\mathcal{G}, T, T)$ where T consists of one edge with two vertices, the corresponding groups being G_1, G_2, and K. By hypothesis, $G_2 = \pi(\mathcal{H}, X, S)$ where all vertex and edge groups of H are finite. Lemmas 1 and 2 give the result.

<u>Corollary. Let $G = G_1 *_K G_2$, where K is finite. If $1 \vee G_1$ and $1 \vee G_2$, then $1 \vee G$.</u>

<u>Proof</u>. Immediate from lemma 4 and the corollary to lemma 3.

<u>Remark</u>. Let $G = \langle G_1, t_i; A_i^{t_i} = B_i \rangle$ where each A_i is finite. If $1 \vee G_1$ then $1 \vee G$. This follows from lemmas 1 and 2 as $G = \pi(\mathcal{G}, X, T)$ where X has one vertex only, and one edge for each index i.

<u>Lemma 5. Let G, H be groups with $G \cap H = K$. If $K \vee G$, then $H \vee H *_K G$.</u>

<u>Proof</u>. We may assume $G = \pi(\mathcal{G}, X, T)$ with $K = G_v$ and all other vertex and edge groups finite. Then $H *_K G = \pi(\mathcal{H}, X, T)$ where $H_v = H$ and $H_w = G_w$ for all other vertices w, $H_e = G_e$ for all edges, whence the result.

Serre's main results show that a group is the fundamental group of a graph of groups iff it acts on a tree. This immediately leads to subgroups theorems from which the next lemmas may be deduced.

Lemma 6. Let $1vG$ and $H \leq G$. Then $1vH$.

Proof. Serre's theorems give immediately that $1vG$ iff G acts on a tree in such a way that the stabiliser of each vertex is finite. The lemma is now clear.

Lemma 7. Let $1vG$ and $H, K \leq G$. If $H\, v\, G$, then $K \cap H\, v\, K$.

Proof. Let $G = \pi(\mathcal{G}, X, T)$, $H = G_v$ with all other vertex and edge groups finite Serre's theorem gives at once $K = \pi(\mathcal{H}, Y, S)$ where the vertex groups of $\mathcal{H}$ consist of $K \cap H$ together with other subgroups of G, and the edge groups of $\mathcal{H}$ are finite (being subgroups of conjugates of the edge-groups of $\mathcal{G}$). By lemma 6, $1\, v\, H_y$ for any vertex y of Y. Lemmas 1 and 2 now give the result.

2. Finitely Generated Groups

For completeness we begin with Solitar's results.

Let $G \in \Phi\mathcal{F}$ be finitely generated. Bu proposition 2.1 (and example 2) of [1], G has at least two ends. Then Stallings' structure theorem (theorem 3.1 and the exercise on p.31 of [1]) tells us that either $G = G_1 *_K G_2$ with K finite, or G is the HNN group $\langle G_1, x;\, K^x = L \rangle$ with K finite. In either case $|G : G_1| = \infty$.

The next lemma, due to Solitar [5], allows us to use induction.

Lemma 8. Let G be either an amalgamated free product $G_1 *_K G_2$ with K finite or an HNN group $\langle G_1, x;\, K^x = L \rangle$ with K finite. If F is a finitely generated free subgroup of G, either $F \leq G_1$ or rank $(F \cap G_1) <$ rank F.

Proof. Plainly F has trivial intersection with any conjugate of K. Hence the subgroup theorems [2, 3, 4] show that $F \cap G_1$ is a free factor of F, giving the result.

<u>Theorem 1 (Solitar). If $G \in \Phi\mathcal{F}$ is finitely generated, then $1 v G$.</u>

<u>Proof</u>. We know that G is either an amalgamated free product $G_1 *_K G_2$ with K finite or an HNN group $<G_1, x; K^x = L>$. By the corollary to lemma 4 and the following remark we need only show $1 v G_1$ (and $1 v G_2$). But by lemma 8, if F is a finitely generated free subgroup of G with $|G : F| < \infty$, then $F \cap G_1$ will be a free subgroup of G_1 with $|G_1 : F \cap G_1| < \infty$ and rank $(F \cap G_1) <$ rank F. So $1 v G_1$ by induction (on the rank of free subgroups of finite index).

<u>Proposition 1. Let B be an infinite subgroup of the finitely generated group G. There is a subset E of G with $\phi \neq E \neq G$ such that $Eb = E$ for all $b \in B$ and $Eg \,\Delta\, E$ finite for all $g \in G$ (Δ denotes symmetric difference) iff either $G = X *_K Y$ with K finite or $G = <X, t; K^t = L>$ with K finite and with $B \leq X$ in either case.</u>

<u>Remarks</u>. If G is as above, G has infinitely many ends.

2. If B is a finite subgroup of G and G has at least two ends, the same result holds using the structure theorem and elementary results on HNN groups and amalgamated free products.

<u>Proof</u>. If G is as stated, the existence of E is easy.

Suppose such an E exists. When G is torsion-free the result is proposition 3.3 of [1]. If we attempted to follow the proof of that result we would run into combinatorial difficulties which could probably be resolved. However it seems easier to generalise Oxley's proof of Stallings' structure theorem. The full details will not be given, but I indicate how the proof in [1] must be modified.

The set E is almost invariant and consequently has finite coboundary δE. Call E a <u>B-minimal</u> set if E is almost invariant, $Eb = E$ for all $b \in B$,

$\phi \neq E \neq G$, and δE has as few elements as possible subject to the above.

Part (iii) of lemma 3.2 of [1] shows that there exists a B-minimal set E with $1 \varepsilon E$ such that if E_1 is B-minimal with $1 \varepsilon E_1 \subseteq E$ then $E_1 = E$.

Part (i) of the same lemma now shows that if E_2 is any B-minimal set at least one of $E \cap E_2$, $E^* \cap E_2$, $E \cap E_2^*$, $E^* \cap E_2^*$ (where $*$ denotes complement) will be finite. Since B is infinite and these four sets are B-invariant, one of these sets must be empty.

Hence the subgroups K, H and subset E_1, defined in the proof of theorem 3.1 of [1] now have the properties

$$K = \{g;\ gE = E\},\ H = \{g;\ gE = E \text{ or } gE = E^*\},$$

$$E_1 = \{g;\ gE \subseteq E \text{ or } gE^* \subseteq E\},$$

$$E_1^* \cup H = \{g;\ gE \subseteq E^* \text{ or } gE^* \subseteq E^*\}.$$

Take any $x \varepsilon E$. Then, by lemma 2.8 of [1], for all but finitely many elements g of E we have either $gE \subsetneq E - \{a\}$ or $gE \supsetneq E^* \cup \{a\}$. Hence $K \cap E$ must be finite. Similarly $K \cap E^*$ is finite, so K is finite (note that we made no assumption on thenumber of ends of G).

A trivial modification (interchanging the roles of G_1 and G_2 and altering the notation) of the proof now gives either $G = X *_K Y$ or $G = \langle X, t;\ K^t = L \rangle$ where $X = \{x \varepsilon G;\ (E_1 - (H - K))x = E_1 - (H - K)\}$. Thus to prove the proposition it is enough to show $E = E_1 - (H - K)$.

As $1 \varepsilon E$, if $g \varepsilon E$ we cannot have $gE \subseteq E^*$ while if $g \varepsilon E^*$ we cannot have $gE \subsetneq E$. In particular $E \cap (H - K) = \phi$. Also $H \subsetneq E_1$, trivially.

Let $g \varepsilon E$. We have just seen that $gE \subsetneq E^*$ is impossible. If $gE^* \subsetneq E^*$, then $gE \supsetneq E$ and $1 \varepsilon g^{-1} E \subsetneq E$. By the choice of E we must have $g^{-1}E = E$

since $g^{-1}E$ is B-minimal. Thus $g \in K \subseteq E_1$, and we have $E \subseteq E_1$.

Let $g \in E_1$, $g \notin E$. As we cannot have $gE \subseteq E$ we must have $gE^* \subseteq E$ so $gE \supseteq E^*$ and $1 \in g^{-1} E^* \subseteq E$. As above this gives $g^{-1}E^* = E$ so $g \in H - K$. Thus $E_1 \leq E \cup (H - K)$, and we have $E_1 - (H - K) = E$, as required.

Corollary. Let G be finitely generated with $1vG$. Let B be a subgroup of G such that there exists a subset E of G, infinite with infinite complement, with $Eg \Delta E$ finite for all $g \in G$ and $Eb = E$ for all $b \in B$. Then there is a finitely generated subgroup X of G with $B \leq X$ and $X \vee G$.

Proof. From the proposition (and remark 2) we have $B \leq X$ with either $G = X *_K Y$ or $G = \langle X, t; K^t = L \rangle$ and K finite. This requires X finitely generated as G is. In the second case lemmas 1 and 2 give $X \vee G$ immediately while in the first case they give $X \vee G$ provided $1vY$, which follows from lemma 6.

3. Countable Groups

We begin with a lemma which generalises a result due to Higman (lemma 6.1 of [11]).

Lemma 9. Let $G \in \Phi\mathcal{F}$. Then the following are equivalent:

(i) $1 \vee G$;

(ii) if $G_1 \leq G_2 \leq \ldots$ is an increasing sequence of finitely generated subgroups of G such that no G_i is contained in a proper vertex of G_{i+1}, then the sequence is ultimately constant;

(iii) for any finitely generated subgroup H of G there is a finitely generated subgroup $K \geq H$ such that $K \vee X$ for any finitely generated $X \geq K$.

Proof. (i)$\Rightarrow$(ii). Let $\bigcup G_n = H$. By lemma 6, $1 \vee H$. There is a finitely generated group K with $G_1 \leq K$ and $K \vee H$ (for we may write $H = \pi(\mathcal{G}, X, T)$ where all vertex groups are finite, and take a finite subgraph Y for which $Y \cap T$ is a maximal tree and $G_1 \leq \pi(\mathcal{G}|Y, Y, Y \cap T)$, and this latter is seen to be a vertex of H by contracting Y).

By lemma 7, $G_{n+1} \cap K \vee G_{n+1}$ for all n. If $G_n \leq K$ we have $G_n \leq G_{n+1} \cap K$ so by hypothesis we must have $G_{n+1} \leq K$. Hence $G_n \leq K$, all n, so $K = H$.

Then H is finitely generated, so $H = G_n$ for large n.

(ii)$\Rightarrow$(iii). Let H_1 be any finitely generated subgroup of G. If $H_1 \vee X$ for every finitely generated subgroup $X \geq H_1$ we take $K = H_1$. If not take $H_2 \geq H_1$ with H_2 finitely generated, H_1 not a vertex of H_2 and such that H_2 has a free subgroup of finite index with smallest possible rank subject to the conditions above on H_2.

As H_2 is finitely generated, any vertex of H_2 can be exhibited by a finite graph. Induction from lemma 8 shows that any proper vertex of H_2 has a free subgroup of finite index whose rank is less than the rank of the corresponding subgroup of H_2. So, by choice of H_2, we see that H_1 is not contained in any proper vertex of H_2.

If $H_2 \vee X$ for every finitely generated $X \geq H_2$ we take $K = H_2$. If not, as above we find a finitely generated $H_3 > H_2$ with H_2 not contained in any proper vertex of H_3. Thus there must be a subgroup K as required, since otherwise we would get a strictly increasing sequence $H_1 < H_2 < H_3 < \dots$, contradicting (ii).

(iii)$\Rightarrow$(i). Let G be generated by $g_1, g_2, \dots$. Let $K_0 = \{1\}$. Since (iii) holds we may define, inductively finitely generated subgroups K_n of G such that $\langle K_{n-1}, g_n \rangle \leq K_n$ and such that $K_n \vee X$ for any finitely generated $X \geq K_n$. In particular, $K_n \vee K_{n+1}$ for all n.

Lemma 3 now gives $1 \vee G$.

Theorem 2. If $G \in \Phi\mathcal{F}\ell$ is countable, $1 \vee G$.

Proof. If not there exists a strictly increasing sequence of finitely generated subgroups G_n such that no G_n is contained in a proper vertex of G_{n+1}. Then G_2 is infinite, since G_2 finite gives $G_1 \vee G_2$. Thus, renumbering, we may assume G_1 infinite.

As already rmarked, $cd_Q G = 1$. Now Swan's proof (proposition 6.2 of [13]) may be followed exactly, starting from the corollary to proposition 1.

3. Uncountable Groups

Denote by I_G the augmentation ideal in QG, i.e. $\ker(QG \to G)$. The following results were proved in [1].

(i) if $G = \langle S \rangle$, I_G is QG-generated by $\{s - 1\}$ (lemma 4.2);

(ii) let $H \leq K \leq G$; if $I_H K$ is a QK-summand of I_K then $I_H G$ is a QG-summand of $I_K G$ (lemma 4.3);

(iii) let $G_i \leq G$, $i = 0, 1, 2, 3$; if $G_3 = G_1 *_{G_0} G_2$ then $I_{G_1} G \cap I_{G_2} G = I_{G_0} G$ and $I_{G_3} G = I_{G_1} G + I_{G_2} G$ (theorem 4.7).

Lemma 10. Let H be a finite subgroup of G. Then $I_H G$ is a summand of QG.

Proof. Let $H = \{h_1, \ldots, h_n\}$. Then $u = \frac{1}{n}\sum_1^n (1 - h_i) \in I_H$, and $u(h_j - 1) = h_j - 1$, $j = 1, \ldots, n$. Thus $x \to ux$ is a QG-homomorphism from QG to $I_H G$ which is the identity on $I_H G$.

Lemma 11. Let G be the HNN group $\langle H, t_i ; A_i^{t_i} = A_i^{\alpha_i} \rangle$, where α_i is a monomorphism of the subgroup A_i of H into H. Let M be a QG-module, f a QG-homomorphism $I_H G \to M$ and u_i elements of M. Then f extends to a homomorphism $I_G \to M$ sending $t_i - 1$ to u_i iff

$$u_i(a_i^{\alpha_i} - 1) = ((a_i - 1)f))t_i - (a_i^{\alpha_i} - 1)f$$

for each i and each $a_i \in A_i$.

Proof. A QG-homomorphism $\phi : I_G \to M$ can be identified with a group homomorphism Φ from G to the semi-direct product of M and G, where $g\Phi = ((g - 1)\phi, g)$. The formula follows immediately as the condition that there is such a group homomorphism sending h to $((h - 1)f, h)$ for $h \in H$ and t_i to (u_i, t_i). Note that this lemma applies to the augmentation ideal over any ring.

Lemma 12. Let $G = \langle H, t; A^t = A^\alpha \rangle$. Then $I_G = I_H G + (t - 1)QG$, and $I_H G \cap (t - 1)QG$ is the submodule M generated by the elements $(t - 1)(a^\alpha - 1) = (a - 1)t - (a^\alpha - 1)$, all $a \in A$.

Proof. As $G = \langle H, t \rangle$, $I_G = I_H G + (t - 1)QG$, and plainly $I_H G \cap (t - 1)QG \supseteq M$. It is enough to show that for any QG-module N, homomorphisms $I_H G \to N$, $(t - 1)QG \to N$ which coincide on M can be extended to a homomorphism $I_G \to N$ (we need only take the projection $I_H G \to I_H G/M$ and the zero map $(t - 1)QG \to I_H G/M$). This follows at once from lemma 11 and the definition of M.

Corollary (a) Let $G = \langle H, t; A^t = A^\alpha \rangle$ with A finite. Then M is a summand of $(t - 1)QG$ and $I_H G$ a summand of I_G.

(b) Let $G = H *_A K$, with A finite. Then $I_H G$ is a summand of I_G.

Proof (a) $(t-1)QG$ is free with basis $t - 1$, so there is an isomorphism $(t - 1)QG \to QG$ mapping M to $I_A \alpha G$. By lemma 10 we have M a summand of $(t - 1)QG$, and lemma 12 now gives $I_H G$ a summand of I_G.

(b) If $G = H *_A K$, $I_A G$ will be a summand of $I_K G$ by lemma 10 and (ii) above, and the result follows by (iii) above.

Theorem 3. If $H \vee G$, then $I_H G$ is a summand of I_G.

Proof. Let $G = \pi(\mathcal{G}, X, T)$ with $G_v = H$ and all other vertex and edge-groups finite.

Suppose first that every edge has v as a vertex. Then G is the amalgamated free product of groups K_e with H amalgamated, where $K_e = \pi(\mathcal{G}|e, e)$. By the corollary to lemma 12 and (ii) above we can write, for any edge, $I_{K_e}G = I_HG \oplus M_e$ for some module M_e. By (i) above we have $I_G = \sum I_{K_e}G = I_HG + (\sum M_e)$, and we find I_G is the direct sum of I_HG and the M_e by an easy induction from (iii) above.

Now let T_n be the subtree of T consisting of all vertices at distance at most n from v, together with the corresponding edges. Let $L = \pi(\mathcal{G}|T, T, T)$ and $L_n = \pi(\mathcal{G}|T_n, T_n, T_n)$. The previous paragraph shows that $I_{L_n}L_{n+1}$ is a QL_{n+1}-summand of $I_{L_{n+1}}$ (since we may contract T_n). Thus $I_{L_n}G$ is a QG-summand of $I_{L_{n+1}}G$, so I_HG is a QG-summand of I_LG, since $H = L_0$, $L = \bigcup L_n$.

Finally I_LG is a summand of I_G by the first paragraph, since we may contract T.

An obvious modification of this proof gives the following.

<u>Theorem 4. Let 1vG. Let R be a ring with unity such that the order of any finite subgroup of G is invertible in R. Then $cd_RG \leq 1$.</u>

<u>Proposition 2. Let H be a finitely generated subgroup of G such that I_HG is a summand of I_G. If 1vG, then H v G.</u>

<u>Proof</u>. Let $G = \pi(\mathcal{G}, X, T)$, all vertex and edge-groups finite. We can find a finite subgraph Y such that $Y \cap T$ is a maximal tree of Y and $H \leq \pi(\mathcal{G}|Y, Y, Y \cap T) = L$, say. Then L v G (contracting Y) and L is finitely generated. Also $L \in \Phi\mathcal{F}$ by Solitar's converse to his conjecture. (In this case it is easy to show $L \in \Phi\mathcal{F}$. By the subgroup theorem of Serre it is enough to find a finite homomorph of L for which the vertex groups map monomorphically. For then the kernel will be free. Finite induction reduces the problem to the cases of an amalgamated free product of finite groups or an HNN group with finite base. The first of these is well-known, and the second has a similar proof to the first.)

Among all finitely generated groups K with $H \leq K$, $K \in \Phi\mathcal{F}$ and $K \vee G$ choose K so that it has a free subgroup of finite index with minimal rank. By lemmas 4.4 and 4.6 of [1], we see first that $I_H K$ is a summand of I_K and then that either $H = K$ (using 4.1 of [1]) or K has a subset E with $\phi \neq E \neq K$, $Ek \Delta E$ finite for all $k \in K$, and $Eh = E$ for all $h \in H$.

If $H = K$ we have $H \vee G$ as required. If not, proposition 1 shows that either $K = K_1 *_A K_2$ or $K = \langle K_1, t; A^t = B \rangle$ where A is finite, and where $H \leq K_1$. As $1 \vee G$ we have $1 \vee K_2$ so $K_1 \vee K$ in either case. So $K_1 \vee G$ and K_1 is finitely generated. By lemma 8, this contradicts the choice of K.

<u>Conjecture 1. Let H be a subgroup of the countable group G such that $I_H G$ is a summand of I_G. If $G \in \Phi\mathcal{F}$ then $H \vee G$.</u>

<u>Conjecture 2. Let H be a subgroup of the countable group G such that $I_H G$ is a summand of I_G. If $1 \vee G$ then $H \vee G$.</u>

By theorem 2, conjecture 2 implies conjecture 1.

<u>Lemma 13. Let $H \leq G$ with $I_H G$ a summand of I_G, and $G = \langle H \cup S \rangle$ for some countable set S. If $G \in \Phi\mathcal{F}$ and conjecture 1 holds, or $1 \vee G$ and conjecture 2 holds, then $H \vee G$.</u>

<u>Proof.</u> As in proposition 7.4 of [1], there exist subgroups L of G, K of H with $G = L *_K H$, $I_K L$ a summand of I_L and L countable. By the conjectures $K \vee L$, so by lemma 5 $H \vee G$.

<u>Lemma 14. Let $G \in \Phi\mathcal{F}$ and $H \leq G$. Then $I_H G$ is a summand of I_G iff $G *_H G \in \Phi\mathcal{F}$.</u>

<u>Proof.</u> Let $K = G *_H G$. Then we have an exact sequence $0 \to I_H K \to I_G K \oplus I_G K \to I_K \to 0$ where the first map is $u \to (u, -u)$ and the second map is the sum of the two inclusions. Applying the automorphism $(u, v) \to (u, u+v)$ of $I_G K \oplus I_G K$ we see there is an exact sequence

$$0 \to I_H K \to I_G K \oplus I_G K \to I_K \to 0$$

where the first map is $u \to (u, 0)$.

Consequently if I_K is projective, $I_H K$ will be a QK-summand of $I_G K$, so that $I_H G$ will be a QG-summand of I_G (since $I_H G$ is plainly a QG-summand of $I_H K$).

Conversely by lemma 4.3 of [1], if $I_H G$ is a QG-summand of I_G we have $I_H K$ a QK-summand of $I_G K$, and $I_G K$ will be QK-projective as I_G is QG-projective. Hence I_K will be a QK-projective, i.e. $cd_Q K \leq 1$.

Thus $K \in \Phi\mathcal{F}$ if it has a torsion-free subgroup of finite index. Since any element of finite order lies in a conjugate of one of the factors, it is enough to find a finite homomorph of K for which the kernel contains no torsion element of either copy of G. We map K to G by identifying the two copies of G, and then map G to a suitable finite group (as $G \in \Phi\mathcal{F}$) to get the result.

Theorem 4. Let $H \leq G$ with $I_H G$ a summand of I_G. If $G \in \Phi\mathcal{F}$ and conjecture 1 holds or 1vG and conjecture 2 holds then H v G.

Theorem 5. If conjecture 1 holds, then $G \in \Phi\mathcal{F}$ implies 1vG.

Proof. Theorem 5 follows from theorem 4, taking $H = \{1\}$.

We follow the proof of theorem D of [1], starting with $cd_Q G = 1$. As in that proof we obtain subgroups G_α of G indexed by the ordinal numbers $\leq \lambda$, with $G_0 = H$, $G_\lambda = G$, $G_\beta = \bigcup_{\alpha<\beta} G_\alpha$ for limit ordinals β, $G_{\alpha+1}$ generated by G_α and a countable set, and $I_{G_\alpha} G_{\alpha+1}$ a summand of $I_{G_{\alpha+1}}$.

By lemma 13, we have (assuming conjecture 1) G_α v $G_{\alpha+1}$ for all α, and the theorem follows from lemma 3.

References

1. Cohen, Daniel E., Groups of Cohomological Dimension One, Springer Lecture Notes 245.

2. Karrass, A. and Solitar, D., The subgroups of a free product of two groups with an amalgamated subgroup, Trans.Amer.Math.Soc. 150 (1970), 227-255.

3.

Subgroups of HNN groups and groups with one defining relation, Can.J.Math. 23 (1971), 627-643.

4. Serre, J. P., Springer Lecture Notes (to appear).

5. Solitar, D., Finitely generated groups with a free subgroup of finite index, J.Aust.Math.Soc., (to appear).

6. Stallings, J., On torsion-free groups with infinitely many ends, Ann. of Math. 88 (1968), 312-334.

7. Swan, R. G., Groups of cohomological dimension one, J.Alg. 12 (1969), 585-610.

FREE SUBGROUPS OF LINEAR GROUPS

John D. Dixon

§1. Introduction

My object is to give an exposition of the main result of [3] which may be stated as follows.

THEOREM (Tits)

Let G be a linear group over a field F. Then either G contains a noncyclic free subgroup or else $G/\text{Solv } G$ is a periodic linear group over F.

In particular, if either G is finitely generated or F has characteristic 0, and G has no noncyclic free subgroup, then G is solvable by finite.

Remarks

When we say G is a linear group over F we mean G has an isomorphic representation as a group of linear transformations of a finite dimensional vector space over the (commutative) field F; equivalently, G is isomorphic to a subgroup of the general linear group $GL(n, F)$ and then n is called the degree of the representation. A linear group G always has a unique maximal normal solvable subgroup (see Lemma 1 below) and we denote this by Solv G.

Our exposition of Tits' theorem follows the general plan of the proof in [3] but differs in several details. By concentrating on the one result we can omit many of the calculations and deeper applications of the theory of algebraic groups which Tits uses. Included in [3] is a list of some of the interesting consequences of this theorem.

§2 Some known results

We shall simplify the exposition by first isolating some of the theorems on which we shall base the proof.

Lemma 1. (Zassenhaus-Mal'cev)

If G is a linear group, then any locally solvable subgroup is solvable; hence it has a unique maximal normal solvable subgroup (the solvable radical of G) which we denote by Solv G.

Proof. See [2] Theorem 6.2B or [4] Corollary 3.8.

We shall need the elementary properties of the Zariski topology with reference to linear groups (see, for example, [2] Chapter 8 or [4] Chapter 5). In particular, the following is needed.

Lemma 2.

Let G be a subgroup of $GL(n, F)$. Then the connected component G^0 of 1 (in the Zariski topology) is a normal subgroup of finite index in G. Moreover G^0 is contained in every closed subgroup of finite index in G.

Proof, See [2] Theorem 8.5 or [4] Lemmas 5.2 and 5.3.

Remark

If F' is a field extension of F, then the topology induced on G by the Zariski topology relative to F' is the same as that induced by the Zariski topology relative to F.

Lemma 3.

Let G be a subgroup of $GL(n, F)$ and H a normal subgroup of G. If H is closed (in the Zariski topology) in G, then G/H is a linear group over F although not necessarily of degree n. Specifically there is a rational representation $f: G \to GL(m, F)$ for some integer m with $\ker f = H$ (f is continuous in terms of the Zariski topologies).

Proof. See [4] Theorem 6.4.

Lemma 4. (Wehrfritz)

Suppose G is a linear group over F and that each finitely generated subgroup is solvable by finite. Then $G/\mathrm{Solv}\, G$ is a periodic linear group over F.

Proof. See [5] Lemma 2.

Lemma 5 (Jordan-Schur)

Suppose that G is a periodic linear group over a field F of characteristic 0. Then G has a normal abelian subgroup A of finite index. (In fact the index $|G:A|$ can be bounded by a function only depending on the degree n of G.)

Proof See [2] Theorem 9.5 or [4] Theorem 9.4.

Finally we have the following modest generalization of a theorem of Schur (compare with §2 of [3]).

Lemma 6. (Schur)

Let G be a finitely generated subgroup of $GL(n, F)$ and let H be a subgroup of G.

(i) If H is periodic, then H is finite.

(ii) If H is completely reducible, and each $x \in H$ has all its eigenvalues roots of unity (that is, a suitable power of x is unipotent), then H is finite.

Proof.

The proof of (i) is essentially the same as that of Theorem 9.2 of [2]. The proof there also shows that there exists an integer N (depending only on G) such that in (ii) x^N is unipotent for all $x \in H$. By Corollary 2.8C of [2] there is no loss in generality in supposing that the irreducible components of H are absolutely irreducible. But since all eigenvalues of elements in H are N th roots of unity, each irreducible component of H only yields a finite number of different trace values. Therefore by Burnside's theorem (Theorem 2.8A of [2]) each of these irreducible components is finite. Therefore H itself is a finite group. This proves (ii).

§3 Initial reductions in the proof

Now consider Tit's theorem. First observe that the final assertion is an immediate corollary of the main assertion of the theorem because of Lemmas 6 and 5. Secondly, Lemma 4 shows that to prove the main assertion it is enough to prove that each finitely generated subgroup of G is solvable by finite, or else G contains a noncyclic free subgroup.

Thus without loss in generality we can restrict ourselves to proving:

(*) Let G be a finitely generated linear group over a field F and suppose G is not solvable by finite. Then G contains a noncyclic free subgroup.

The steps of the proof of (*) will be: choose a suitable field and suitable generators for the free subgroup; then construct a new representation for G in which we can prove that the generators generate a free subgroup. It is helpful to recall that if the elements aN and bN are free generators of a free subgroup in a factor group G/N, then a and b are free generators of a free subgroup of G (use the universal property of free groups!)

In the following we shall consider a finitely generated subgroup G of GL(n, F) where G is not solvable by finite. It is readily seen that we can assume with no loss of generality that G is completely reducible as well (see Theorem 2.4 of [2]). Since G^0 is normal and of finite index in G (see Lemma 2), G^0 is completely reducible by Clifford's theorem (Theorem 2.2 of [2]) and finitely generated since G is. Thus we may replace G by G^0 and assume that G is connected in the Zariski topology. Let G' denote the derived group.

(A) <u>There exists $x \in G'$ with at least one eigenvalue not a root of unity</u>.

Since G is completely reducible, so is G' by Clifford's theorem. However G' cannot be finite, for otherwise $G/C_G(G')$ would be finite (it may be embedded in Aut G'). But the isomorphism $C_G(G')/Z(G') \simeq C_G(G')G'/G' \subseteq G/G'$ shows that $C_G(G')$ is solvable, and so then G would be solvable by finite contrary to hypothesis. Hence G' is infinite and so (A) follows from Lemma 6(ii).

The first major step of the proof is construction of a suitable normed field over which we take our representations. Let E_0 be the prime subfield of F and let E be any finitely generated subfield of F containing all entries from a set of generators of G (so $G \subseteq GL(n, E)$).

(B) For any $\xi \in E$ which is not a root of unity there exists an extension field $\hat{E}$ of E with a norm (= multiplicative valuation) $|\;|$ such that $\hat{E}$ is locally compact with respect to $|\;|$ and $|\xi| \neq 1$.

(i) (char E = 0). Since E is finitely generated and the complex field $\mathbb{C}$ is algebraically closed with infinite transcendence degree, there are embeddings $E \to \mathbb{C}$. Moreover, an embedding can be chosen so that the image $\bar{\xi}$ of ξ has absolute value $\neq 1$; this is obvious when ξ is transcendental, and when ξ is algebraic we use the fact that in $\mathbb{C}$ only roots of unity have all their algebraic conjugates of absolute value 1. Since $\mathbb{C}$ is locally compact under the topology induced by the absolute value, it follows that we can take $\hat{E}, |\;| \simeq \mathbb{C}, |\;|$.

(ii) (char F = p $\neq$ 0). Now each element of E algebraic over E_0 is already a root of unity, so ξ is necessarily transcendental. Let $\xi = \xi_1, \ldots, \xi_k$ be a transcendence basis of E over E_0, put $D = E_0[\xi_1, \ldots, \xi_k]$ let P be the ideal generated by $\xi_1, \ldots \xi_k$ in D. Then we have a uniquely defined valuation $|\;|$ on $E_0(\xi_1, \ldots \xi_k)$ such that if $\alpha \in D$, then $|\alpha| = 2^{-i}$ when $\alpha \in P^i \setminus P^{i+1}$ $(i = 0, 1, \ldots)$. This valuation may be extended to the algebraic extension E of

$E_0(\xi_1,\ldots,\xi_k)$ and then we may take the completion $\hat{E}, ||$ of $E, ||$ (we continue to use the same symbol $||$ for the norm). Since the quotient field D/P is finite, and the norm $||$ is discrete, $\hat{E}, ||$ is a locally compact field; indeed, $\hat{D} = \{\alpha \in \hat{E} \mid |\alpha| \leq 1\}$ is a compact neighbourhood of 0, and in general $\beta + \hat{D}$ is a compact neighbourhood of β (see [1] p. 49-50). By construction $|\xi| = 2^{-1} \neq 1$ and so (B) is proved in this case too.

We now take a fixed element $x \in G'$ with an eigenvalue ξ which is not a root of unity (see (A)), and then choose E in (B) to contain all eigenvalues of x as well as all entries of the matrices in G. By (B) we can find a locally compact field $\hat{E}, ||$ containing E in which $|\xi| \neq 1$. Note that $\det x = 1$ because $x \in G'$ ($\det a^{-1}b^{-1}ab = 1$!) and so $|\xi| \neq 1$ implies not all eigenvalues of x have the same norm.

To simplify notation we shall now replace $\hat{E}, ||$ by $F, ||$. As a subgroup of $GL(n, F)$ G may now no longer be completely reducible but G is still connected (see the Remark after Lemma 2).

§4. Constructing new representations

Our object is to show that a power of x and a suitable conjugate generate a free subgroup. To prove this it is necessary to construct a new representation of G in which x and x^{-1} have dominant eigenvalues; then multiplication by powers of x will "attract" vectors towards the corresponding eigenspaces.

(C) <u>We may suppose that x has a unique simple eigenvalue λ of largest norm.</u> We saw at the end of §3 that not all eigenvalues of x have the same norm. Therefore suppose x has (counting multiplicities) r eigenvalues

$\lambda_1,\ldots,\lambda_r$ of norm δ and the other $n-r$ eigenvalues of norm $< \delta$, (so $\delta > 1$ and $1 \leq r \leq n-1$ by the remarks at the end of §3).

Now consider the free exterior algebra over F with basis elements $e_1,\ldots e_n$ (thus $e_i \wedge e_j = -e_j \wedge e_i$ and $e_i \wedge e_i = 0$). Let V be the F-subspace of this algebra with the basis of r-mononials $e_{i_1} \wedge \ldots \wedge e_{i_r}$ $(i_1 < \ldots < i_r)$; so $\dim_F V = \binom{n}{r}$. We construct a representation of G on V by defining

$$(e_{i_1} \wedge .. \wedge e_{i_r})\, a = \sum_j \alpha_{i_1 j} e_j \wedge .. \wedge \sum_j \alpha_{i_r j} e_j$$

when $a = [\alpha_{ij}] \in G$. If we define $\rho(a)$ for the matrix of this linear transformation over the basis of r-monomials of V, then we have a rational representation $\rho : G \to (\binom{n}{r}, F)$ for G. Since the eigenvalues of $\rho(a)$ are just all possible products of r eigenvalues of a, $\rho(x)$ has a unique simple eigenvalue $\lambda = \lambda_1 \cdots \lambda_r$ of norm δ^r and all other eigenvalues of norm $< \delta^r$.

Note that in (C) the eigenspace V_1 for $\rho(x)$ corresponding to λ is not a $\rho(G)$-subspace. For otherwise we should have a representation of G of degree 1 on V_1; but the kernel of any representation of degree 1 contains G' whilst clearly x is not in the kernel of this particular representation.

Since ρ is a rational mapping the image $\rho(G)$ is connected since G is. We shall replace $\rho(G)$ by G and $\binom{n}{r}$ by n. Our new G is a finitely generated connected subgroup of $GL(n, F)$ and $x \in G'$ has a unique eigenvalue λ of largest norm. Let V be the underlying vector space of n-row vectors over F. We shall write

$$V = V' \oplus V''$$

where V' is the eigenspace of x corresponding to λ and V'' is a complementary x-subspace.

(D) <u>There exists $b \in G$ such that</u> $V' \cap V''b = V'' \cap V'b = V' \cap V'b = 0$

Indeed put $L = \{z \in G \mid V'z = V'\}$

$$M = \{z \in G \mid V'z \subseteq V'' \text{ or } V'z^{-1} \subseteq V''\} .$$

Then L and M are closed subsets of G (write down the polynomial identities defining them!) Since $x \in L$, $L \neq \emptyset$; and the remark following (C) shows that $L \neq G$. But G is connected and $L \cap M = \emptyset$, therefore $L \cup M \neq G$. Hence we can take any $b \in G$, $b \notin L \cup M$.

Our final representations σ for G is constructed by defining

$$\sigma(a) = \begin{bmatrix} a & 0 \\ 0 & (a^{-1})^T \end{bmatrix}$$

where $(a^{-1})^T$ is the transpose of a^{-1}. The underlying vector space may be written in the form $W = V \oplus \bar{V}$ where $\bar{V}$ is a copy of V on which the lower block of $\sigma(a)$ acts. Once again we replace $\sigma(G)$ by G and $2n$ by n. The element x now has unique eigenvalues λ and $1/\lambda$ of greatest and least norms (the eigenvalues of $(a^{-1})^T$ are the reciprocals of those of a). We can write

$$W = V_\ell \oplus U_s \oplus U_0$$

where V_ℓ, V_s are the 1-dimensional eigenspaces for x for the eigenvalues λ and $1/\lambda$, respectively, and V_0 is a complementary x-subspace. Similarly using (D) and writing $y = b^{-1}xb$ we have

$$W = U_\ell \oplus U_s \oplus U_0$$

where U_ℓ, U_s are eigenspaces for y for the eigenvalues λ and $1/\lambda$, respectively, and U_0 is a complementary y-subspace. Because of (D) we have:

V_ℓ and V_s are both disjoint from $U_\ell + U_0$ and $U_s + U_0$;

U_ℓ and U_s are both disjoint from $V_\ell + V_0$ and $V_s + V_0$.

We now apply the topological properties of $F, ||$. W inherits the product topology from F and as such is a locally compact (Hausdorff) space. Since W is finite dimensional all subspaces of W are closed. Hence we have the following.

(E) Let v_ℓ, v_s, u_ℓ, u_s be nonzero vectors spanning V_ℓ, V_s, U_ℓ, U_s, respectively. Then there exist compact subsets A and B of W such that

(i) A is a neighbourhood of v_ℓ and v_s; and B is a neighbourhood of u_ℓ and u_s

(ii) $a\alpha = b\beta$ with $\alpha, \beta \in F$ and $a \in A$, $b \in B \Rightarrow a\alpha = 0 = b\beta$.

(iii) $A \cap (U_s + U_0) = A \cap (U_\ell + U_0) = \emptyset$

$B \cap (V_s + V_0) = B \cap (V_\ell + V_0) = \emptyset$.

§5. <u>Conclusion of the proof</u>

It remains to show that suitably large powers of x and y generate a free subgroup of rank 2. This rests on the following property.

(F) For all sufficiently large integers m:

$Bx^m \cup Bx^{-m} \subseteq AS$

$Ay^m \cup Ay^{-m} \subseteq BS$

(Here S denotes the set of all scalars in $GL(n, F)$ and AS, BS are simply the scalar multiples of elements in A,B, respectively).

Consider the relation $Bx^m \subseteq AS$. First note that since the restriction of $\lambda^{-1}x$ to $V^* = V_s + V_0$ has all its eigenvalues with norm strictly less than 1, its powers converge to 0. Now $W = V_\ell \oplus V^*$

and so if $w \in W \backslash V^*$, then $w = \alpha v_\ell + v^*$ for some $\alpha \neq 0$ in F and some $v^* \in V^*$; and therefore $wx^m/\alpha\lambda^m \to v_\ell$ as $m \to \infty$. Thus if $w \notin V^*$, then $wx^m \in AS$ for all sufficiently large m. Since $B \cap V^* = \emptyset$ by (E) and B is compact, there exists m_0 such that $wx^m \in AS$ for all $w \in B$ whenever $m \geq m_0$; that is, $Bx^m \subseteq AS$ for all sufficiently large m. The proof of the other three relations is analogous.

The proof is now completed as follows. Take m so large that (F) is satisfied. We claim $\langle x^m, y^m\rangle S/S$ is free of rank 2. Otherwise there would exist a word of the type

$$z = x^{mk_1} y^{m\ell_1} \ldots x^{mk_s} y^{m\ell_s} \in S$$

where all $k_1, \ell_1, \ldots k_s, \ell_s$ are nonzero integers, and $s \geq 1$. But then (F) shows that

$$v_\ell x^{mk_1} \in AS\ , \quad v_\ell x^{mk_1} y^{m\ell_1} \in BS, \ \ldots \ v_\ell x^{mk_1} y^{m\ell_1} \ldots y^{m\ell_s} \in BS.$$

By (E) (ii) this contradicts the condition $v_\ell z \in v_\ell S \subseteq AS$. Thus $\langle x^m, y^m\rangle S/S$ is free and so from the observation in §3 $\langle x^m, y^m\rangle$ itself is free. This proves Tits' theorem.

REFERENCES

1. J.W.S. Cassels and A. Fröhlich (eds), Algebraic Number Theory, Washington D.C.: Thompson Book Co. 1967.

2. J.D. Dixon, The Structure of Linear Groups. London: Van Nostrand-Reinhold Co. 1971.

3. J. Tits, "Free subgroups in linear groups", J. Algebra 20 (1972) 250-270.

4. B.A.F. Wehrfritz, Infinite Linear Groups. London: Queen Mary College Mathematics Notes 1969.

5. ______________, "2-generator conditions in linear groups", Archiv Math. 22(1971) 237-240.

On Groups of Exponent Four IV

C. C. Edmunds and N. D. Gupta

INTRODUCTION. In [9] Wright proved that the nilpotency class of an n-generator group of exponent four is at most $3n - 1$. Recently Gupta and Quintana [2] have shown that an improvement in Wright's bound to $3n - 3$ for any n would imply the solvability of groups of exponent four in general and thus verify a conjecture of G. Higman [4] and M. Hall (oral communication). In this note our main observation is that the Higman-Hall conjecture has a very strong impact on the class of an n-generator group. More specifically we prove the following:

Theorem A. If G is a solvable group of exponent four, then there is an integer k (depending only on the solvability length of G) such that every n-generator subgroup of G is nilpotent of class at most $n + k$.

It follows from the proof of Theorem 3.2 of Tobin [8], that if G is a group of exponent four and $N \lhd G^2$ ($= \mathrm{gp}\{g^2 : g \in G\}$), then $[N, 4G] \leq [N, (G^2)^2][N, N]$. From this it is easy to deduce that if G is a solvable group of exponent four, then G^2 is nilpotent of class depending only on the solvability length of G. Thus our Theorem A is a consequence of the following:

Theorem B. If G is a group of exponent four with G^2 nilpotent of class r, then every n-generator subgroup of G is nilpotent of class at most $n + k$ where $k = r$ if $n \geq 2r + 2$ and $k = [4(r+1)/3] + 3$ if $n < 2r + 2$.

LEMMAS. In this section we list several lemmas which will be required for the proof of Theorem B. The reader is referred to Hanna Neumann's book [6] for any unexplained notation.

For the remainder of the paper the symbol * will be used to indicate the presence of an entry in a commutator. For instance, writing $c = [*, *, \cdots, x, \cdots]$ would indicate that the entry x avoids the first two positions of c. Note that a repeated * in a commutator does not necessarily assume the same value for each of its occurrences.

Let $\gamma_n(G)$ denote the $n^{\underline{th}}$ term of the lower central series of G.

Lemma 1. If G is a group of exponent four, the following congruence relations hold.

(1.1) $[a,b]^2 \equiv 1 \bmod \gamma_4(G)$

(1.2) $[a,b,c] \equiv [a;b,c][a,c,b] \bmod \gamma_4(G)$

(1.3) $[x,a;b,c] \equiv [x,b;a,c][x,c;a,b] \bmod \gamma_5(G)$

(1.4) $[x,a,a,y] \equiv [y,a,a,x][x,a,y,y][y,a,x,x]$
$\times[x,a;y,a][x,y;y,a][x,y;x,a] \bmod \gamma_5(G)$

(1.5) $[x,a;y,b;c] \equiv [x,b;y,c;a][x,c;y,a;b] \bmod \gamma_6(G)$

(1.6) $[x,y,a,z,b] \equiv [x,y,z,a,b][x,z,a,b,y][x,z,b,a,y]$
$\times[x,b,a,y,z]\ [x,b,y,a,z] \bmod \gamma_6(G)$

(1.7) $[x,a,a,y,a] \equiv [x,a,y,a,a] \bmod \gamma_6(G)$

(1.8) $[x,y,z,a,a,a] \equiv [x,y,a,a,a,z] \equiv 1 \bmod \gamma_7(G)$

(1.9) $[x,a,a,a,y,y] \equiv 1 \bmod \gamma_7(G)$

(1.10) $[x,a,a,a;y,z] \equiv 1 \bmod \gamma_7(G)$.

Proof. Congruences (1.1), (1.2), (1.3), (1.5), (1.6), and (1.7) are found in Wright's paper [9]. Congruence (1.2) is the well-known Jacobi congruence; (1.6) follows from the congruence at the bottom of page 388 of [9] using x instead of $[x, y]$, replacing z by y, w by z, a by b, b by a and rearranging the factors; and (1.7) follows from the second congruence on page 391 of [9] using x instead of $[x, y]$ and replacing z by y. Congruences (1.4) and (1.8) are recent observations of M. Hall [3].

By (1.1) $[x, y^2] \equiv [x, y, y] \bmod \gamma_4(G)$. To prove (1.9) we replace x by $[x, a]$ and y by y^2 in (1.4) and use the above congruence to get

$$\begin{aligned}[x,a,a,a,y,y] &\equiv [y^2,a^2;x,a][x,a^2;y^2,a] \\ &\equiv [x,y^2;a,a^2] \qquad \text{by (1.1) and (1.3)} \\ &\equiv 1 \bmod \gamma_7(G) .\end{aligned}$$

Since $\gamma_2(G) \leq G^2$, (1.10) is an easy consequence of (1.9).

In the following lemma every commutator is in $\gamma_m(G)$, and those entries not specifically written remain in the same order under each congruence.

Lemma 2. If G is a group of exponent four, the following congruence relations hold modulo $\gamma_{m+1}(G)$ $(m \geq 6)$.

(2.1) $[*, \cdots, a,a, \cdots] \equiv [*, \cdots, a^2, \cdots]$

(2.2) $[\cdots, x,y,z, \cdots] \equiv [\cdots, x;y,z; \cdots][\cdots, x,z,y, \cdots]$

(2.3) $[*, \cdots, a,a,x,a, \cdots] \equiv [*, \cdots, a,x,a,a, \cdots]$

(2.4) $[*,*, \cdots, a,a,a, \cdots] \equiv 1$

(2.5) $[*,a,a,a, \cdots, x^2, \cdots] \equiv 1$

(2.6) $[*,a,a,a, \cdots; x,y; \cdots] \equiv 1$

(2.7) $[\cdots, a, \cdots, a, \cdots, a, \cdots, a, \cdots] \equiv 1$.

Proof. Congruences (2.1), (2.2), (2.3), and (2.4) follow directly from (1.1), (1.2), (1.7), and (1.8) respectively. Proving (2.5) and (2.6) simultaneously we let

$$c \equiv [*,a,a,a,z_1, \cdots, z_n,d, \cdots] \in \gamma_m(G)$$

where $d = x^2$ or $[x, y]$ and proceed by induction on n . If $n = 0$, then $c \equiv 1$ by (1.9), (1.10), and (2.1). If $n > 0$, then by (2.2)

$$\begin{aligned} c &\equiv [*,a,a,a,z_1, \cdots, z_{n-1};z_n,d; \cdots][*,a,a,a,z_1, \cdots, z_{n-1},d,z_n,\cdots] \\ &\equiv 1 \text{ by the induction hypothesis.} \end{aligned}$$

Congruence (2.7) is a result of Wright [9].

In the following lemma every commutator is in $\gamma_m(G)$, and those entries not specifically written are present in some permuted form in each factor.

<u>Lemma 3.</u> If G is a group of exponent four, the following congruence relations hold modulo $\gamma_{m+1}(G)(m \geq 6)$.

(3.1) $[\cdots, a, \cdots]$ is congruent to a product of commutators of the type $[a, \cdots]$.

(3.2) $[*,*, \cdots, a, \cdots, a, \cdots]$ is congruent to a product of commutators of the types

$$[*,*, \cdots, a^2, \cdots] \text{ and } [\cdots, a,*,a] .$$

(3.3) $[\cdots, w,*,x, \cdots, y,a,*,b,z, \cdots]$ is congruent to a product of commutators of the types

$$[*, \cdots, a,b, \cdots], [*, \cdots, b,a, \cdots] ,$$

and the single commutator

$$[\cdots, w,a,*,b,x, \cdots, y,*,z, \cdots]$$

or $$[\cdots, w,b,*,a,x, \cdots, y,*,z, \cdots]$$

where all the entries of the single commutator, except possibly x and y, remain in the same order.

(3.4) $[*,*, \cdots, a, \cdots, a, \cdots, a, \cdots]$ is congruent to a product of commutators of the type

$$[\cdots, a^2,*,a] .$$

Proof. Congruences (3.1) and (3.2) are results of Wright [9]. To prove (3.3) we let

$$c = [\cdots, w,*,x_1, \cdots, x_n,a,*,b,y, \cdots]$$

and proceed by an easy induction on n using (1.6). Congruence (3.4) follows from Wright's Lemma 2 (p. 390, [9]), (2.4), and (2.1).

PROOF OF THEOREM B. Let $G = gp\{x_1, \cdots, x_n\}$ be a group of exponent four with G^2 nilpotent of class r. It suffices to prove that $\gamma_m(G) \leq \gamma_{m+1}(G)\,\gamma_{r+1}(G^2)$ where $m = n + r + 1$ if $n \geq 2r + 2$ and $m = n + [4(r + 1)/3] + 4$ if $n < 2r + 2$. We will show that if $c = [y_1, \cdots, y_m]$ is a commutator such that $\{y_1, \cdots, y_m\} \subseteq \{x_1, \cdots, x_n\}$, then modulo $\gamma_{m+1}(G)$ c is congruent to a product of commutators each of which contains at least $r + 1$ entries from G^2. It is not difficult to see that we need only consider the case when $\{y_1, \cdots, y_m\} = \{x_1, \cdots, x_n\}$.

Now we give two routines (D and T) which will be applied repeatedly to yield the final result. We say that x_i is a single (double, triple) entry of the commutator c if x_i occurs exactly once (twice, three times) as an entry of c.

Routine D. The input for this routine is a commutator

$$c' = [*,*, \cdots, x, \cdots, x, \cdots] \in \gamma_m(G) \quad (m \geq 6)$$

which has a double entry x and another entry y ($\neq x$) which we will refer to as a distinguished entry. The output is a factorization of c' modulo $\alpha_{m+1}(G)$ into a product of commutators of the types

$$[*, \cdots, x^2, \cdots] ,$$

$$[*, \cdots, x;x,y; \cdots] ,$$

and

$$[y,x,*,x, \cdots]$$

where now y , the leftmost x , and the rightmost x will be used as distinguished entries of the respective types in the next application of Routine D.

The routine proceeds by applying (3.2) modulo $\gamma_{m+1}(G)$, to get c' congruent to a product of commutators of the types

$$[*, \cdots, x^2, \cdots] \quad \text{and} \quad [\cdots, x,*,x] .$$

By (3.2) and (2.1) this product is congruent to a product of commutators of the types

$$[*, \cdots, x^2, \cdots], [*, \cdots, x,y,x, \cdots] ,$$

and

$$[y,x,*,x, \cdots] .$$

Finally by (2.2) we get c' congruent to a product of commutators of the types

$$[*, \cdots, x^2, \cdots] , [*, \cdots, x;x,y; \cdots] ,$$

and

$$[y,x,*,x, \cdots] .$$

This is the desired output.

Henceforth, the entries x^2, $[x, y]$, and $[y, x]$ $(\in G^2)$ will be considered unavailable for use as input in further applications of Routines D or T.

Routine T. The input for this routine is a commutator

$$c' = [*,*, \cdots, x, \cdots, x, \cdots, x, \cdots] \in \gamma_m(G) \quad (m \geq 6)$$

which has a triple entry x and a distinguished entry y $(\neq x)$. The output is a factorization of c' modulo $\gamma_{m+1}(G)$ into a product of commutators of the types

$$[*, \cdots, x^2, \cdots; x,y; \cdots],$$

$$[*, \cdots; x,y; \cdots, x^2, \cdots] ,$$

and

$$[y,x, \cdots, x^2, \cdots] .$$

This routine begins with an application of (3.4) modulo $\gamma_{m+1}(G)$ to get c' congruent to a product of commutators of the type

$$[\cdots, x^2,*,x] .$$

By (3.3) this is congruent to a product of commutators of the types

$$[*, \cdots, x^2,x, \cdots] , \quad [*, \cdots, x,x^2, \cdots] ,$$

$$[*, \cdots, x^2,y,x, \cdots] , \quad [*, \cdots, x,y,x^2, \cdots] ,$$

$$[y,x^2,*,x, \cdots], \text{ and } [y,x,*,x^2, \cdots] .$$

By (2.1), (2.2), (2.3), and (2.4) this is congruent to a product of commutators of the types

$$[*,x,x,x, \cdots, y, \cdots] ,$$

$$[y,x,x^2, \cdots] ,$$

$$[*, \cdots, x^2;x,y; \cdots] ,$$

and $$[y,x,*,x^2, \cdots] .$$

Thus it suffices to show that commutators of the form

$$d = [a,x,x,x,z, \cdots, y, \cdots]$$

can be factored appropriately. By repeated applications of (2.2) and (2.6) we get

$$d \equiv [a,x,x,x,y,z, \cdots] .$$

Modulo $\gamma_7(G)$

$$\begin{aligned} [a,x,x,x,y,z] &\equiv [a,x^2;x,y;z][a,x,x,y,x,z] \quad \text{by (2.1) and (2.2)} \\ &\equiv [a,x^2;x,y;z][a,x,y,x,x,z] \quad \text{by (1.7)} \\ &\equiv [a,x^2;x,y;z][a;x,y;x,x,z][a,y,x,x,x,z] \quad \text{by (2.2)} \\ &\equiv [a,x^2;x,y;z][a;x,y;x^2,z] \quad \text{by (1.8) and (2.1)} . \end{aligned}$$

Thus modulo $\gamma_{m+1}(G)$

$$d \equiv [a,x^2;x,y; \cdots][a;x,y;x^2, \cdots] ,$$

which is the desired factorization.

As in Routine D x^2, $[x, y]$, and $[y, x]$ are now unavailable for use as input in further applications of Routines D or T .

We note that by the nature of the congruences used in each of these routines, at each step the entries not used as input are present in some permuted form in each factor.

By (2.7) we may assume that no entry of $c(=[y_1, \cdots, y_m])$ occurs more than three times; and since c contains n distinct entries in m ($> n$) positions, c must have some double or triple entries. Let s, d, and t be the number of x_i's which are single, double, and triple entries of c respectively. Clearly we have $n = s + d + t$ and $m = s + 2d + 3t$.

Case I. If $n \geq 2r + 2$, we choose $m = n + r + 1$. Thus $r + 1 = m - n = (s + 2d + 3t) - (s + d + t) = d + 2t$. Therefore $0 \leq t \leq [(r+1)/2]$, $0 \leq d \leq r + 1$, and $s = n - d - t > n - (r + 1) - [(r + 1)/2] = n - [3(r+1)/2] \geq (2r + 2) - [3(r + 1)/2] \geq [(r + 1)/2] \geq t$.

If $d \neq 0$ we begin using Routine D with $c' = c$, x any double entry avoiding the first two positions of c, and y any single entry of c. The distinguished entry which is produced as output from this application of the routine will be used the next time a distinguished entry is needed. We continue this process until no double entries avoid the first two positions of the commutators which we have produced. Thus we arrive at a product of commutators in $\gamma_m(G) \cap \gamma_d(G^2)$ and commutators of the types

$$[x, *, \cdots, x, \cdots] \in \gamma_m(G) \cap \gamma_{d-1}(G^2)$$

and
$$[x, y, \cdots, x, \cdots, y, \cdots] \in \gamma_m(G) \cap \gamma_{d-2}(G^2)$$

where each type contains either a single or a distinguished entry.

To a commutator of the second type we apply (3.1) in order to shift this single or distinguished entry into the first position. We then apply Routine D using the double entry which has become available and the first entry as input. This reduces each commutator of the second type to a product of commutators of the first type.

Upon applying (3.1) to the single or distinguished entry, z, of a commutator of the first type we arrive at a product of commutators of the types

$$[z,*, \cdots, x, \cdots, x, \cdots]$$

and $$[z,x, \cdots, x, \cdots] \quad .$$

We apply Routine D to the first type and leave the second type as it is considering $[z, x]$ as one entry.

Thus c is congruent to a product of commutators in $\gamma_m(G) \cap \gamma_d(G^2)$ each of which contains t triple entries and either s single entries or $s - 1$ single entries and one distinguished entry. In either case we apply Routine T repeatedly using single or distinguished entries and triple entries as input. Eventually we arrive at a product of commutators in $\gamma_m(G) \cap \gamma_{d + 2t}(G^2)$ and commutators of the types

$$[x,*, \cdots, x, \cdots, x, \cdots] \in \gamma_m(G) \cap \gamma_{d+2(t-1)}(G^2) \quad ,$$

and $$[w,x, \cdots, w, \cdots, w, \cdots, x, \cdots, x, \cdots] \in \gamma_m(G) \cap \gamma_{d+2(t-2)}(G^2)$$

where both types contain single or distinguished entries y and z different from w and x and each commutator of the second type contains an additional single entry. Commutators of the second type can be reduced to products of commutators in $\gamma_m(G) \cap \gamma_{d+2t}(G^2)$ and commutators of the first type by shifting the additional single entry to the first position and applying Routine T to the triple entries which now avoid the first two positions. With a commutator of the first type, we shift y into the first position and apply Routine T to the three x's and y if possible. Otherwise we have commutators of the type

$$[y,x, \cdots, x, \cdots, x, \cdots, z, \cdots] \in \gamma_m(G) \cap \gamma_{d+2(t-1)}(G^2) .$$

We then consider $[y, x]$ $(\in G^2)$ as one entry and apply Routine D to the two rightmost x's and z.

Thus modulo $\gamma_{m+1}(G)$, c is congruent to a product of commutators which are in $\gamma_{d+2t}(G^2) \le \gamma_{r+1}(G^2)$.

If $d = 0$ we omit Routine D and proceed as above.

Case II. If $n < 2r + 2$ we apply Routine D to c as in Case I using one occurrence of a double entry as the first distinguished entry. After $d - 1$ applications of Routine D we arrive at a product of commutators in $\gamma_m(G) \cap \gamma_{d-1}(G^2)$ each of which contains two entries, y and z, which will be available for future use as distinguished entries.

Next we apply Routine T as in Case I, but instead of using single entries as input we use triple entries. We do this systematically using the three occurrences of a triple entry as the distinguished entries in three consecutive applications of Routine T. This yields a ratio of six new entries from G^2 in the output for all the occurrences of four triple entries used as input for three applications of Routine T.

Exhausting this process we arrive at a product of commutators in $\gamma_m(G) \cap \gamma_{(d-1)+t_1}(G^2)$ $(t_1 = [3(t-1)/2])$ and commutators in $\gamma_m(G) \cap \gamma_{(d-1)+t_2}(G^2)$ $(t_2 = [3(t-2)/2] = [3t/2] - 3)$. The latter commutators contain three occurrences of a triple entry x, and the distinguished entries y and z if $d \neq 0$. Using (3.1) we shift y to the first position and apply Routine T to y and x if x avoids the first two positions. Otherwise we arrive at commutators of the type

$$[y,x, \cdots, x, \cdots, x, \cdots, z, \cdots] \in \gamma_m(G) \cap \gamma_{(d-1)+t_2}(G^2) .$$

We then consider $[y, x]$ as one entry and apply Routine D to the two rightmost x's and z .

If $d = 0$, we apply only Routine T and arrive at a product of commutators in $\gamma_m(G) \cap \gamma_{[3t/2]}(G^2)$ and commutators in $\gamma_m(G) \cap \gamma_{t_2}(G^2)$ of the types

$$[x,y, \cdots, x, \cdots, x, \cdots]$$

and $$[x,*, \cdots, y, \cdots, x, \cdots, x, \cdots] .$$

We leave a commutator of the first type as it is considering $[x, y]$ as one entry. With a commutator of the second type we apply Routine D to the two rightmost x's and y .

Thus for all d we get c congruent to a product of commutators in $\gamma_m(G) \cap \gamma_{d+t_2+1}(G^2)$.

Since $m = n +([4(r + 1)/3]+ 3)+ 1$, we get $2t + d = [4(r + 1)/3]+ 4$,

and therefore

$$t = \frac{1}{2}[4(r + 1)/3]+ 2 - \frac{1}{2}d \ .$$

Thus $$\frac{3}{2}t + d - 2 = \frac{3}{4}[4(r + 1)\ 3] + \frac{1}{4}d + 1 \geq r + 1 \ .$$

This completes the proof of Theorem B.

<u>CONCLUDING REMARKS.</u> We remark that Theorem B yields new information about third Engel groups of exponent four. It follows from Theorem 2 of Gupta and Weston [1] that a third Engel group G of exponent four satisfies the law

$$[[x_1^2, x_2^2; x_3^2, x_4^2],[x_5^2, x_6^2; x_7^2, x_8^2],[x_9^2, x_{10}^2]] = 1 \ .$$

Since $[G^2, G^2] = (G^2)^2$, it follows by the result of Tobin mentioned in the introduction letting $N = \delta_2(G^2)$ that $\gamma_6((G^2)^2) \leq \delta_3(G^2)$, where $\delta_n(G)$ is the $n^{\underline{th}}$ term of the derived series of G. Thus $\gamma_7((G^2)^2) \leq [\delta_3(G^2), [G^2, G^2]] = \{1\}$. By six more applications of Tobin's result we obtain $\gamma_{26}(G^2) \leq \gamma_7([G^2, G^2]) = \{1\}$. Together with Theorem B, this implies that for large enough n the class of an n-generator subgroup of a third Engel group of exponent four is at most $n + 25$. This improves the results of Ivanjuta [5], $2n$, and Quintana and Wright [7], $2n + 1$. It follows from a recent result of M.F. Newman that the exact class is $n + 2$.

In the case of groups of exponent four in general the authors know of examples which show that for $r = 1, 2,$ and 3 the bound $n + r$ is best possible provided n is large enough.

REFERENCES.

[1] N. D. Gupta and K. W. Weston, "On Groups of Exponent Four," J. Algebra 17(1971), 59-66.

[2] N. D. Gupta and R. B. Quintana, "On Groups of Exponent Four III," Proc. Amer. Math. Soc. 33(1972), 15-19.

[3] M. Hall, "Notes on Groups of Exponent Four," (to appear).

[4] G. Higman, "The Orders of Relatively Free Groups," Proc. Internat. Conf. Theory of Groups, Canberra (1965), 153-165.

[5] Ī. D. Ivanjuta, "Certain Groups of Exponent Four," Dopovīdī Akad. Nauk Ukraïn. RSR Ser. A(1969), 787-790, 860.

[6] H. Neumann, Varieties of Groups, Springer-Verlag, Berlin (1967).

[7] R. B. Quintana and C. R. B. Wright, "On Some Groups of Exponent Four," Pacific J. Math. (to appear).

[8] S. Tobin, "On a Theorem of Baer and Higman," Canad. J. Math. 8(1956), 263-270.

[9] C. R. B. Wright, "On the Nilpotency Class of a Group of Exponent Four," Pacific J. Math. 11(1961), 387-394.

The Higman Theorem for $E^n(A)$ Computable Groups

R. W. Gatterdam[1]

Introduction

The concept of a computable group was introduced by Rabin [10] and extended by Cannonito and Gatterdam to $E^n(A)$ computable groups in [1], [5] and [2]. In essence a countable group is computable if it can be encoded onto a decidable subset of the natural numbers in such a way that, relative to the encoding, the group operations of multiplication and inversion are computable. Such an encoding is called an index. The class of $E^n(A)$ functions, for $A \subset N$, consists of functions from N^k to N derived from the initial functions for the Grzegorczyk hierarchy, E^n, together with the characteristic function of A by substitution and limited recursion ([8], [11]). Basically, $E^2(A)$ contains the zero function, projections, the successor, addition, multiplication, the characteristic function for A and all those functions obtained from them by substitution and recursion provided the recursion has a bounding function already in $E^2(A)$. An unbounded recursion leads from class $E^n(A)$ to $E^{n+1}(A)$. An $E^n(A)$ group then is a group having an $E^n(A)$ computable index.

The concept of an $E^n(A)$ group is closely related to the word problem for finitely generated (f.g.) groups. In fact, every f.g. group G can be given an $E^3(A)$ index for some $A \subset N$ in such a way that the decidability of the word problem for a f.g. presentation of G is the same as that for A. Any presentation of G yields an index for G called a standard index having computability that of the word problem. It is known that if one f.g. standard index of G is $E^n(A)$ for $n \geq 3$ then all f.g. standard indices are $E^n(A)$ and so we may talk of $E^n(A)$ standard groups. It is also known that a f.g. $E^n(A)$ group for $n \geq 3$ is $E^{n+1}(A)$ standard. For details see [2].

Constructions

In [2] the constructions of free product with amalgamation and strong Britton extension are studied. It is shown that if G_1, G_2 are $E^n(A)$ groups for $n\geq 3$, $H_1<G_1$, $H_2<G_2$ are $E^n(A)$ decidable and $\phi: H_1 \rightarrow H_2$ is an isomorphism such that ϕ and ϕ^{-1} are $E^n(A)$ computable, (considered as functions from the index of H_1 to that of H_2) then $G_1 *_\phi G_2$ is $E^{n+1}(A)$ computable. Also, if G is $E^n(A)$ for $n\geq 3$, $H_1<G$, $H_2<G$ are $E^n(A)$ decidable and $\phi: H_1 \rightarrow H_2$ is an isomorphism such that ϕ and ϕ^{-1} are $E^n(A)$ computable, then $G_\phi = \langle G, t; tht^{-1} = \phi(h)\ \forall h \in H_1 \rangle$ is $E^{n+1}(A)$ computable. For G_1, G_2 and G $E^n(A)$ standard for $n\geq 4$ and for ϕ and ϕ^{-1} $E^{n-1}(A)$ computable, it is known that $G_1 *_\phi G_2$ and G_ϕ are $E^n(A)$ standard.

In [2] by a technique of Cannonito [1], it is shown that for $n\geq 4$ and $A\subseteq \mathbb{N}$ there exist f.g. $E^n(A)$ standard groups not $E^{n-1}(A)$ standard and such that if $E^m(B)$ standard, B is $E^n(A)$ decidable.

Results

The following is a compilation of the main results from [2] and [7].

Theorem 1: Every $E^n(A)$ group for $n\geq 3$ can be embedded in a two generator $E^{n+1}(A)$ standard group. Every $E^n(A)$ standard group for $n\geq 4$ can be embedded in a two generator $E^n(A)$ standard group.

Theorem 1 is the Higman Neumann Neumann theorem for $E^n(A)$ groups and is proved in the standard fashion using the constructions described above.

Theorem 2: Every f.g., $E^n(A)$ standard group for $n\geq 4$ and A recursively enumerable can be embedded in a finitely presented $E^n(A)$ standard group.

Theorem 2 is the Higman theorem, [9], for $E^n(A)$ groups. The technique of proof is based on Schoenfield [12] as modified in [5]. It is shown in [5] that the Clapham result, [3], [4], also can be obtained by this technique.

The embeddings of theorems 1 and 2 are $E^n(A)$ embeddings in that not only are they $E^n(A)$ computable as homomorphisms, but in addition the image of the embedded group is $E^n(A)$ decidable as a subgroup of the embedding group. Theorems 1 and 2 together yield:

Corollary 3: Every $E^n(A)$ standard group for $n \geq 4$ and A recursively enumerable can be embedded in a finitely presented $E^n(A)$ standard group.

Together with Cannonito's technique for constructing f.g. $E^n(A)$ standard groups we get:

Corollary 4: For $n \geq 4$ and A recursively enumerable there exists a finitely presented group G which is $E^n(A)$ standard and not $E^{n-1}(A)$ standard. If G is $E^m(B)$ standard then B is $E^n(A)$ decidable.

And in particular, in answer to a question raised in [1]:

Corollary 5: For $n \geq 4$ there exist E^n standard groups which are not E^{n-1} standard.

AMS 1969 subject classifications. Primary 02F47, 20F05, 20F10. Secondary 02F35, 20E30.

Key words and phrases. Computable groups, group presentation, word problem, embeddings of groups, Higman theorem, computable function.

---

[1]University of Wisconsin Graduate School Project 120455 (Grant).

References

[1] F. B. Cannonito, Hierarchies of computable groups and the word problem, Journal of symbolic logic 31 (1966) 376-392.

[2] F. B. Cannonito and R. W. Gatterdam, The computability of group constructions, part I, Proceedings of the Irvine conference on decision problems in group theory (North Holland in preparation).

[3] C. R. J. Clapham, Finitely presented groups with word problem of arbitrary degrees of insolubility, Proceedings of the London Mathematical Society 14 (1966) 633-676.

[4] C. R. J. Clapham, An embedding theorem for finitely generated groups, Proceedings of the London Mathematical Society 17 (1967) 419-430.

[5] R. W. Gatterdam, Embeddings of primitive recursive computable groups, doctoral dissertation, University of California, Irvine, 1970. Submitted to Annals of mathematical logic.

[6] R. W. Gatterdam, The Higman theorem for primitive recursive groups--a preliminary report, Proceedings of the Irvine conference on decision problems in group theory (North Holland in preparation).

[7] R. W. Gatterdam, The computability of group constructions part II, to appear.

[8] A. Grzegorczyk, Some classes of recursive functions, Rozprawy Mathmetyczne 4 (1953) 46 pp.

[9] G. Higman, Subgroups of finitely presented groups, Proceedings of the Royal Society, A 262 (1961) 455-475.

[10] M. O. Rabin, Computable algebra, general theory and theory of computable fields, Transactions of the American Mathematical Society 95 (1960) 341-360.

[11] R. W. Ritchie, Classes of recursive functions based on Ackerman's function, mimeographed lecture notes, University of Washington, 1963.

[12] J. R. Schoenfield, Mathematical Logic (Addison Wesley 1967).

The University of Wisconsin-Parkside

RING THEORETIC METHODS AND FINITENESS CONDITIONS IN INFINITE SOLUBLE GROUP THEORY

K. W. Gruenberg

In recent years, ring theory has come to play an increasingly important part in the study of infinite groups. Probably the most striking progress in this direction was made by Philip Hall in his three marvellous papers [1], [2], [3].

I should like to speak to you(*) in general terms about how ring theory can be brought into play for studying structural questions on groups and to confine the more specific discussion to the ideas in Hall's first paper. The second and third papers of Hall will feature in Dr. Roseblade's lecture.

The basic idea behind the use of ring theory is (necessarily perhaps) exceedingly simple. Suppose G is a group and H/K is an abelian slice of G: here H and K are normal subgroups of G, and K contains H', the commutator subgroup of H. Now G acts by conjugation on H/K and so we obtain a ring homomorphism of $\mathbb{Z}G$, the integral group ring of G, into $\mathrm{End}(H/K)$, the endomorphism ring of H/K. If we know enough about the structure of the image of $\mathbb{Z}G$ in $\mathrm{End}(H/K)$ and about H/K as a G-module, then we may hope to extract information about the way the slice H/K is situated in G.

Here is the simplest relevant illustration. G is a finitely generated metabelian group and G' is to be our slice: $K = 1$, $H = G'$. Since G acts on G' exactly as does G/G', therefore the image, R say, of $\mathbb{Z}G$ in $\mathrm{End}(G')$ is a finitely generated commutative ring. All such rings are noetherian by the Hilbert basis theorem. Now G' is finitely generated as normal subgroup

---

* I am grateful to Martha Smith for drawing my attention to some complete nonsense that appeared towards the end of my lecture. Hopefully none remains in this present version.

of G and so is finitely generated as R-module, whence G' is actually a noetherian R-module. Thus all of commutative noetherian ring theory becomes available for the study of finitely generated metabelian groups. Much structural information about these groups can be proved very simply by using this observation.

As an example let us prove

(1). All finitely generated metabelian groups satisfy the maximal condition on normal subgroups.

Proof. We have to establish that every normal subgroup H of our finitely generated metabelian group G is finitely generated as normal subgroup. Since all subgroups of G/G' are finitely generated, we may suppose $H \leqslant G'$. But then H is an R-submodule of G', with R as in our previous remarks, whence H is finitely generated as R-module. So H is finitely generated as normal subgroup.

To give this type of reasoning wider validity one has to extend the relevant ring theory to a wider, non-commutative, context. This is precisely what Hall was able to do. As a result of this work and J.E. Roseblade's recent splendid completion of it, a number of important results about polycyclic groups carry over to the class of finitely generated abelian-by-polycyclic groups. (We recall that a polycyclic group is a soluble group satisfying Max, the maximal condition on all subgroups.) We shall see (cf. (4) and (5) below) that every finitely generated abelian-by-polycyclic group is in Max-n, the class of groups with maximal condition on all normal subgroups; and every soluble Max-n group, in turn, is easily seen to be finitely generated. Thus we have

(2). Polycyclic $<$ f.g. abelian-by-polycyclic

$<$ Max-n $\cap$ soluble

$<$ f.g. soluble.

The fact that these inequalities are all strict follows from the examples $Z \wr Z$, $(Z \wr Z) \wr Z$, $Z \wr (Z \wr Z)$ respectively, where Z denotes an infinite cyclic group and $\wr$ denotes the usual abstract wreath product. (Cf. [1].)

Perhaps it is fair to conjecture that Hall's papers grew from an attempt to classify finitely generated soluble groups but that they revealed these groups to be so staggeringly diverse that a true understanding of the whole class seems a very long way off. As a matter of fact, at this time, almost no non-trivial properties common to all finitely generated soluble groups have revealed themselves. One exception is the following result which was spotted independently by Robinson [5] and Wehrfritz [6] and which is a direct consequence of Hall's work: if every finite image of a finitely generated soluble group is nilpotent, then the group itself is nilpotent.

Let us return to the ring theory. We considered a situation where a group ring acts as a noetherian ring on an abelian slice. When can this be expected?

If R is a ring, we shall write that R has Max-r, Max-ℓ, Max-2 according as R has the maximal condition on right ideals (is right noetherian), on left ideals (is left noetherian), on two-sided ideals (is noetherian), respectively. If $R = \mathbb{Z}G$ is a group ring, then G will be said to have Max-r, Max-ℓ, Max-2 according as R has these properties.

(3). Within the class of all groups we have

$$\text{Max-r} = \text{Max-}\ell \leqslant \text{Max} \quad \text{and} \quad \text{Max-2} \leqslant \text{Max-n}.$$

The proof, which is very simple, involves a basic correspondence between subgroups and ideals. Recall first that the augmentation ideal of $\mathbb{Z}G$ is the $\mathbb{Z}$-space spanned by all $g - 1$, for g in G. (We shall consistently use small german letters to denote augmentation ideals of groups denoted by the corresponding roman capitals.) If $H \leqslant G$ then $\mathfrak{h}G$ is the right ideal of $\mathbb{Z}G$ generated by $\mathfrak{h}$.

Lemma. If $H \leqslant G$, then $(1 + \mathfrak{h}G) \cap G = H$.

The proof of this is only an exercise. But it yields (3) immediately: by the lemma, $H_1 < H_2$ always implies $\mathfrak{h}_1G < \mathfrak{h}_2G$ and thus Max-r $\leqslant$ Max. Moreover, if H is normal in G, $\mathfrak{h}G$ is automatically a two-sided ideal and therefore we also have Max-2 $\leqslant$ Max-n. Finally, $g \mapsto g^{-1}$ is an additive automorphism of $\mathbb{Z}G$ which gives a bijection of the lattice of right ideals onto that of the left ideals, whence Max-r = Max-ℓ.

We can now prove the proper generalization of (1):

(4). If the finitely generated group E has an abelian normal subgroup so that $G = E/A$ has Max-r, then E has Max-n.

Proof. By (3), E/A has Max and hence we need only prove A is a noetherian G-module. Now

$$A \cong \mathfrak{a}E/\mathfrak{e}\mathfrak{a}, \text{ as } G\text{-modules},$$

by the obvious map: $a \mapsto (1-a) + \mathfrak{e}\mathfrak{a}$. (The inverse is $t(1-a) + \mathfrak{e}\mathfrak{a} \mapsto a$, where t runs through a transversal to the cosets of A in E.) Since E is finitely generated so $\mathfrak{e}$ is finitely generated as E-module and therefore $\mathfrak{e}/\mathfrak{e}\mathfrak{a}$ is finitely generated as G-module. But $\mathbb{Z}G$ is right noetherian and consequently

$\mathfrak{n}/\mathfrak{n}\mathfrak{a}$ is a noetherian module. Thus the submodule $\mathfrak{a}E/\mathfrak{n}\mathfrak{a}$ is also noetherian, as required.

Max-r is an unsatisfactory group theoretic property in the sense that it refers for its definition to a structure outside the group itself. Can we characterise the class Max-r by internal group theoretic properties? This is an interesting but thorny question. For soluble groups the answer is yes and is provided by Hall's generalization of Hilbert's basis theorem in [1] :

(5). Polycyclic = Max-r $\cap$ soluble = Max $\cap$ soluble.

It follows that all polycyclic-by-finite groups have right noetherian group rings. All known groups with Max are polycyclic-by-finite and this has led to the question of whether perhaps these two classes coincide. Since Max-r (by (3) and (5)) is sandwiched between them, it may possibly be worth taking a serious look at the problem: is every group in Max-r necessarily polycyclic-by-finite? It is just possible that this may be more accessible than the similar question with Max in view of the fact that a great deal has been discovered in recent years about noetherian rings.

We recall that if R is a right noetherian ring with no nilpotent ideals - i.e., a semiprime ring - then there exist prime ideals $P_1,\ldots,P_k$ such that $P_1 \cap \ldots \cap P_k = 0$ and each prime ring R/P_i has a right quotient ring which is a complete matrix ring over some division ring (Goldie's fundamental structure theorem).

Suppose G lies in Max-r. By a result of Passman [4], $\mathbb{Z}G$ is semiprime. If we can prove that the image of G in each prime image of $\mathbb{Z}G$ is polycyclic-by-finite, then so will G be. The problem thus reduces to the study of subgroups H of GL(n,D), where D is a division ring and H has Max-r. Note that when D is commutative, then H must indeed be polycyclic-by-finite by by a powerful recent theorem of Tits (cf. Dr. Dixon's lecture).

Returning to soluble groups, we state, for comparison with (5), the following result:

(6). Polycyclic < Max-2 $\cap$ soluble < Max-n $\cap$ soluble.

This strongly suggests that further relations between finiteness conditions on soluble groups and their group rings remain to be discovered. We shall conclude with a discussion of the proof of (6).

Suppose H is normal in G, $S = \mathbb{Z}H$, $R = \mathbb{Z}G$. Then G acts on S as a group of ring automorphisms via conjugation on H. If I is an ideal of S ($I \lhd S$) and $g^{-1}Ig = I$ for all g in G, we shall call I a G-ideal of S and write $I \lhd_G S$. Moreover, we denote the maximal condition on G-ideals by Max-2G.

If $I \lhd_G S$, then IG, the right ideal of R generated by I, is automatically two-sided; and rather obviously $IG \cap S = I$. Hence $I_1 < I_2$ (both being G-ideals of S) implies $I_1G < I_2G$. We therefore have

(7). If S does not have Max-2G, then R does not have Max-2.

We are now in a position to exhibit a soluble group in Max-n but not in Max-2. Let $G = Z_2 \wr Z$, where Z_2 is cyclic on a of order 2 and Z is infinite cyclic on x. Let H be the normal closure of Z_2. Then H is an elementary 2-group with basis a_i, $i \in \mathbb{Z}$, where $a_i = x^{-i}a\, x^i$ and x is the automorphism: $a_i \mapsto a_{i+1}$. Since G is finitely generated metabelian, $G \in$ Max-n (by (1)). To show $G \notin$ Max-2 we use (7) and prove $S = \mathbb{Z}H \notin$ Max-2G.

Let $\alpha_i = a_i - 1$ and $\sigma_k = \alpha_0\alpha_k\alpha_{2k}\cdots\alpha_{(k-1)k}$ for all $k \geq 2$. Denote by I_k the G-ideal in S generated by $\sigma_2,\ldots,\sigma_k$. We assert

$I_{k-1} < I_k$ for all $k \geqslant 3$ (hence $S \notin$ Max-2G). Pick some $k \geqslant 3$ and let φ be the ring homomorphism determined by $H \to H/K$, where K is the subgroup generated by all a_i, with i not divisible by k, and all $a_0^{-1}a_{nk}$, for n in $\mathbb{Z}$. Then H/K is cyclic of order 2 and

$$\alpha_i\varphi = 0 \text{ if } k \text{ does not divide } i,$$
$$\alpha_{nk}\varphi = \alpha_0\varphi \text{ for all } n.$$

Now for any t in $\mathbb{Z}$ and any r in the range $2 \leqslant r < k$,

$$\sigma_r^{x^t} = \alpha_t\alpha_{t+r}\cdots$$

and either k does not divide t or k does not divide $t+r$. Therefore in any case, $(\sigma_r^{x^t})\varphi = 0$, whence $I_{k-1}\varphi = 0$. But $I_k\varphi \neq 0$ because $\sigma_k\varphi = (\alpha_0\varphi)^k$, which is non-zero because $S\varphi = \mathbb{Z}\langle a_0\varphi\rangle$, $(a_0\varphi)^2 = 1$ and $\alpha_0\varphi = a_0\varphi - 1$ is not nilpotent.

To prove the first inequality of (6) we need another extension of Hilbert's basis theorem. Again consider $H \lhd G$, $S = \mathbb{Z}H$, $R = \mathbb{Z}G$. Let M be an S-bimodule which is also a G-module. We call M an S^G-bimodule if $(s_1ms_2)^g = s_1^g m^g s_2^g$ for all s_1, s_2 in S, m in M, g in G. In particular, S and all its G-ideals as well as R itself are S^G-bimodules (with G acting always by conjugation).

If Hx is central in G/H, then Sx $(= xS)$ is also an S^G-bisubmodule of R. The right S-module isomorphism $S \to Sx$ given by $s \mapsto xs$ maps G-ideals onto S^G-bisubmodules and from this we conclude that S has Max-2G if, and only if, Sx has the maximal condition on S^G-bisubmodules. More generally we have the following

<u>Lemma.</u> If $Hx_1, \ldots, Hx_k$ are central in G/H, and S has Max-2G, then $Sx_1 \oplus \ldots \oplus Sx_k$ has the maximal condition on S^G-bisubmodules.

(8). Let $\bar{H} = \langle H, x \rangle$, where Hx is central in G/H and set $\bar{S} = \mathbb{Z}\bar{H}$.

Then S has Max-2G if, and only if, $\bar{S}$ has Max-2G.

Proof. The implication $\Leftarrow$ is trivial and in any case we do not need it. So we discuss only $\Rightarrow$. The proof is effectively the same as that of Hilbert.

If $x^m \in H$, then $\bar{S} = S \oplus Sx \oplus \ldots \oplus Sx^{m-1}$ and we are done by the lemma. So assume $\bar{H}/H$ is infinite. Take $I \trianglelefteq_G \bar{S}$ and let

$$I^{\geqslant 0} = I \cap (\bigoplus_{i \geqslant 0} Sx^i).$$

Then $I^{\geqslant 0}$ generates I as right ideal of $\bar{S}$ and is an S^G-bimodule. The set J of leading coefficients of all the elements of $I^{\geqslant 0}$ is itself a G-ideal of S and so, by hypothesis, is generated by a finite number of elements: say by $a_1,\ldots,a_k$. Let $u_1,\ldots,u_k$ by elements in $I^{\geqslant 0}$ having leading coefficients $a_1,\ldots,a_k$, respectively, and assume them all of the same degree n. Suppose $u_1,\ldots,u_k$ generate the G-ideal $\bar{I}$ of $\bar{S}$. Again applying the lemma, we see that $V = S \oplus Sx \oplus \ldots \oplus Sx^{n-1}$ has maximal condition on S^G-bisubmodules. If $u \in I^{\geqslant 0}$ and has degree ${}^{\circ}u \geqslant n$, then we construct in the expected manner an element u_1 in $I^{\geqslant 0} \cap \bar{I}$ so that ${}^{\circ}(u - u_1) < {}^{\circ}u$. Repeating this procedure as often as necessary we finally arrive at an element u^* in $I^{\geqslant 0} \cap V = W$ with $u^* \equiv u \mod \bar{I}$. Hence $I^{\geqslant 0} \leqslant \bar{I} + W$ and so I is generated as G-ideal by $\bar{I}$ and W. Now $\bar{I}$ is finitely generated as G-ideal by construction, and W is finitely generated as S^G-bimodule (and therefore also as G-ideal of $\bar{S}$) since $W \leqslant V$. Hence I is finitely generated, as required.

An immediate consequence of (8) is

(9). If G/H is a finitely generated nilpotent group, then $\mathbb{Z}H$ has Max-2G if, and only if, $\mathbb{Z}G$ has Max-2.

To complete matters we need

(10). Suppose $H \lhd G$ and that H is the normal closure of a subgroup A in Max-2. If, for each finite subset T of H, there exists g in G so that $T^g \leqslant A$, then $\mathbb{Z}H$ has Max-2G.

Proof. Suppose to the contrary that we can find a sequence $\sigma_1, \sigma_2, \ldots$ in $\mathbb{Z}H$ so that, if J_k is the G-ideal generated by $\sigma_1, \ldots, \sigma_k$, $J_k < J_{k+1}$ for all k. Let σ_i involve the finite subset T_i of H and suppose $T_i^{g_i} \leqslant A$. By Max-2, there exists m so that $\mathbb{Z}A \cap J_m = \mathbb{Z}A \cap J_{m+1} = \ldots$. Now $\sigma_{m+1}^{g_{m+1}}$ lies in $\mathbb{Z}A$, therefore in $\mathbb{Z}A \cap J_m$ and so $\sigma_{m+1} \in J_m$, a patent contradiction.

It is now a simple matter to construct a group tailored to fit (9) and (10) and therefore to provide the example needed to complete (6). We let G be $\langle a, x;\ x^{-1}a\,x = a^2 \rangle$ and H be the normal closure of $\langle a \rangle$. Then G/H is infinite cyclic and $H = \langle a_i\ ;\ a_i = x^i a x^{-i}$ all $i \in \mathbb{Z} \rangle$. Thus H is the group of dyadic rationals and locally cyclic. Every finite subset of H lies in some $\langle a_i \rangle = \langle a_o \rangle^{x^{-i}}$. Consequently $\mathbb{Z}H$ has Max-2G by (10) and then $\mathbb{Z}G$ has Max-2 by (9). But obviously G is not polycyclic.

References

[1] P. Hall, Finiteness conditions for soluble groups, Proc. London Math. Soc. (3) 4 (1954) 419-36.

[2] P. Hall, On the finiteness of certain soluble groups, Proc. London Math. Soc. (3) 9 (1959) 595-622.

[3] P. Hall, The Frattini subgroups of finitely generated groups, Proc. London Math. Soc. (3) 11 (1961) 327-52.

[4] D. S. Passman, Nil ideals in group rings, Mich. Math. J. 9 (1962) 375-84.

[5] D. J. S. Robinson, A theorem on finitely generated hyper-abelian groups, Inventiones Math. 10 (1970) 38-43.

[6] B.A.F. Wehrfritz, Groups of automorphisms of soluble groups, Proc. London Math. Soc. (3) 20 (1970) 101-22.

The Index of an Algebra Automorphism Group

Franklin Haimo[1]

Let M be an infinite set, let Sym M be the symmetric group on M, and let A = A(M; f_t, $t \in T$) be a finitary, universal algebra on M with operations f_t indexed by a set T. If f_t is n-ary (i.e., of rank n) we write $|f_t| = n$. We shall relate the index of the permutation subgroup Aut A in Sym M to the problem of replacing A by a "less complicated" algebra B with Aut A = Aut B. A permutation $\Gamma \in$ Sym M may or may not commute with an operation f_t ($f_t \Gamma = \Gamma f_t$ or $f_t \Gamma \neq \Gamma f_t$) depending upon whether or not $(x_1, \ldots, x_n) f_t \Gamma = (x_1 \Gamma, \ldots, x_n \Gamma) f_t$ for every $(x_1, \ldots, x_n) \in M^n$ where $|f_t| = n$.

Lemma 1. The following are equivalent: (1) Γ commutes with f_t; (2) each member of the coset (Aut A)Γ commutes with f_t; (3) each member of the coset Γ Aut A commutes with f_t.

Theorem 1. If the index n = [Sym M : Aut A] is finite then there exists C = C(M; f), an algebra on M with a single finitary operation f, such that Aut C = Aut A (as permutation groups on M).

Proof. If n = 1, f may be taken to be the identity map from M to M. If n > 1 choose representatives $\Gamma_1, \ldots, \Gamma_{n-1}$ for the n-1 non-trivial, distinct (say, right) cosets of Aut A in Sym M. Since Γ_i is no automorphism of A there exists at least one operation f_i of A such that f_i and Γ do not commute. (There may be repetitions among these n-1 f_i's.) If B = B(M; $f_1, \ldots, f_{n-1}$),

Aut A $\leq$ Aut B. If $\Gamma \in$ Aut B then Γ commutes with each f_i ($i = 1, \ldots, n-1$). By Lemma 1, Γ is excluded from each non-trivial coset of Aut A in Sym M. Hence $\Gamma \in$ Aut A, so that Aut A = Aut B. But a theorem of Gould's [1] allows us to construct C = C(M; f) with Aut B = Aut C. ∎

Recall that Jónsson [2] [3] has identified those subgroups of Sym M that are automorphism groups of finitary algebras on M as the closed subgroups in a certain topology on Sym M, a topology first studied by Karrass and Solitar [4]. If a closed subgroup H of Sym M were to be the automorphism group of no A(M; f_t, $t \in T$, T finite) then the above theorem shows that [Sym M: H] is infinite. We can, however, ease the restriction on H by an analysis of the set O of operations f_t of A. If f_s, $f_t \in O$ we say that f_s and f_t are <u>equivalent</u> ($f_s \sim f_t$)if $|f_s| = n = |f_t|$, and if there exists a permutation λ of the set $\{1, \ldots, n\}$ such that $f_s = \lambda f_t$ in that $(x_1, \ldots, x_n)f_s = (x_{1\lambda}, \ldots, x_{n\lambda})f_t$ for every $(x_1, \ldots, x_n) \in M^n$. The relation $\sim$ is an equivalence; let $[f_t]$ be the equivalence class in O to which f_t belongs. Write $[f_s] \leq [f_t]$ if $n = |f_s| \leq |f_t| = m$, and if there exists at least one map φ from $\{1, \ldots, m\}$ onto $\{1, \ldots, n\}$ such that $f_s = \varphi f_t$. The relation $\leq$ is well defined and partially orders the set O^* of equivalence classes $[f_t]$ under $\sim$. We say that f_s and f_t are <u>comparable</u> if $[f_s] \leq [f_t]$ or if $[f_t] \leq [f_s]$. If $\Gamma \in$ Sym M, if Γ commutes with f_t, and if $[f_s] \leq [f_t]$ then Γ commutes with f_s. In particular, if $f_s \sim f_t$ then both these operations commute with the same permutations.

Let A and B be finitary algebras on M. If the operations of B constitute a subclass of the operations of A, and if Aut A = Aut B we say that B <u>is</u>

__superior to__ A ($A \leq B$), thus introducing a partial order $\leq$ in the class of finitary algebras on M.

__Lemma__ 2. Let $O = U \cup V$ where U and V are non-empty, disjoint subsets of O such that to each $f_s \in U$ there exists at least one $f_t \in V$ with $[f_s] \leq [f_t]$. Then the algebra on M with operations V is superior to the algebra on M with operations O.

__Theorem__ 2. Given a finitary algebra A on the infinite set M, there exists at least one finitary algebra B on M such that (1) $A \leq B$, (2) the set P of operations of B decomposes into the disjoint union of two subsets (not both necessarily non-void) S and W such that (3) if f_s and f_t are comparable in P then both f_s and f_t lie in W, (4) if B' is an algebra on M obtained from B by replacing S by one of its subsets S' of lower cardinality then $B \not\leq B'$, and (5) if B" is an algebra on M obtained from B by replacing W by one of its subsets W' of lower cardinality then $B \not\leq B''$.

__Proof__. By expelling all but one member from each $[f_s]$, replace the algebra A by an algebra A_1 on M in which each $[f_s]$ has only one member, so that $A \leq A_1$. In A_1 we lose no generality if we write $f_s \leq f_t$ in place of $[f_s] \leq [f_t]$. Let O_1 be the set of operations of A_1, and let S_1 be the set of all f_s in O_1 which do not lie in infinite chains. That is, if $f_1 < f_2 < \ldots$ is any infinite chain in O_1 then $f_s \neq f_i$ for $i = 1, 2, \ldots$. Let S_1' be the set of maximal elements of S_1. By Lemma 2, the algebra on M with operations $S_1' \cup (O_1 \setminus S_1)$ is superior to A_1. No two elements of S_1' are comparable, while each element of $O_1 \setminus S_1$

lies in at least one infinite chain. Using the well ordering of the cardinals, one can replace S_1' by one of its subsets S of minimal cardinality without altering the automorphism group of the algebra; likewise, for a subset W of $O_1 \setminus S_1$. ■

Any such B obtained from A as above is called a *rectification of* A, so that, by Theorem 2, any finitary algebra on an infinite set can be rectified in at least one way. The cardinals $|S|$ and $|W|$ are called *rectification cardinals for* A. Since there may be many ways of rectifying A, there may be many such cardinals.

Theorem 3. If A has at least one infinite rectification cardinal then $I = [\text{Sym } M : \text{Aut } A]$ is exceeded by no rectification cardinal for A.

Proof. Suppose that I is smaller than some infinite rectification cardinal for A so that $I < \max(|S|, |W|)$ for some rectification B of A where $B = B(M; S \cup W = P)$. A modification of the proof of Theorem 1 provides an algebra C on M such that $\text{Aut } A = \text{Aut } B = \text{Aut } C$, and $C = C(M; R)$ where $R \subseteq P$ and $|R| \leq I$. If $|R \cap S| = |S|$ and $|R \cap W| = |W|$ then $|R| = |P| = \max(|S|, |W|)$. Since $|R| \leq I < \max(|S|, |W|)$, we have a contradiction. Hence $|R \cap S| < |S|$ or $|R \cap W| < |W|$. If both $|S|$ and $|W|$ are infinite then $R \subseteq P$ and $\text{Aut } B = \text{Aut } C$ contradict the assumption that B rectifies A. If, say, only $|S|$ is infinite and if $|R \cap S| < |S|$ then, again, B could not rectify A. If, however, only $|S|$ is infinite, and if $|R \cap S| = |S|$ then $|R| = |R \cap S| = |S|$. But $|R| \leq I < |S|$, a contradiction. One argues likewise if only $|W|$ is infinite. ■

Example I. Let T be an infinite subset of a set M. On M define a family of unary operations indexed by T, $\{f_t \mid t \in T\}$, by setting $xf_t = t$ for every $x \in M$. Then the algebra $A = A(M; f_t, t \in T)$ is rectified with void W and with $P = S = \{f_t \mid t \in T\}$. By the theorem, $[\mathrm{Sym}\ M ; \mathrm{Aut}\ A] \geq |T|$. Since the ranks of the operations of A are bounded, a result of Gould's [1] shows the existence of an algebra B on M with only one finitary operation such that Aut A = Aut B. The converse of Theorem 1 is accordingly false. If T = M the constructed algebra A has the property that Aut A = 1, consistent with the lower bound $|M|$ on the index.

Example II. Let $(a_1, a_2, \ldots)$ be an infinite sequence of distinct elements $a_i \in M$. Define a sequence $(g_1, g_2, \ldots)$ of operations on M by setting $(x_1, \ldots, x_n)g_n = a_j$ if there are precisely j distinct elements among the n x_i's. Here, $[g_1] < [g_2] < \ldots$, S is void, and $P = W = \{g_1, g_2, \ldots\}$. By the theorem, I is at least countably infinite. Let P' be a subset of P where $P \setminus P'$ is finite. Then the algebra $A' = A'(M; P')$ is likewise a rectification of A, whence A has a countably infinite number of rectifications.

References

[1] M. Gould, Automorphism groups of algebras of finite type (to appear in the Canadian Journal of Mathematics)

[2] B. Jónsson, Algebraic structures with prescribed automorphism groups, Colloq. Math. 19 (1968), 1-4

[3] _____, Topics in universal algebra, Lecture notes in mathematics, Issue 250, Springer -Verlag, 1972

[4] A. Karrass and D. Solitar, Some remarks on the infinite symmetric group, Math. Zeitschr. 66 (1956), 64-69

WASHINGTON UNIVERSITY

ST. LOUIS, MO

Footnote

[1]This work was supported, in part, by National Science Foundation grant number GP-20291.

Notes on Groups of Exponent Four*

Marshall Hall Jr.

California Institute of Technology

1. Introduction

If F_k is the free group generated by k elements $x_1, x_2, \ldots, x_k$, and if F_k^4 is its fully invariant subgroup generated by all fourth powers, then $B(k) = F_k/F_k^4$ is the Burnside group B(k) of exponent 4 with k generators. Sanov [3] has shown that B(k) is finite but little more than that. Much more precise information is Wright's [5,6] result that the $3k^{th}$ term in the lower central series is the identity, $B(k)_{3k} = 1$.

Section 2 gives an explicit finite presentation for B(k). Section 3 gives some general relations on commutators, mostly drawn from Wright. But there is also established a sort of third Engel condition, namely $(x,y,z,a,a,a) \equiv 1 \pmod{G_7}$.

Using the relations of Section 3, commutator relations are found in B(3), including $B(3)_9 = 1$ which show that the order of B(3) is at most 2^{72}.

---

*This research was supported in part by ONR contract NOOO 14-67-A0094-0010

2. Notes on identities in groups of exponent 4

Let $F_k = \langle x_1, \ldots, x_k \rangle$ be the free group generated by $x_1, \ldots, x_k$. The Burnside group B(k) of exponent 4 with k generators is the factor group F_k/F_k^4, where $F_k^4 = \langle y^* | \ y \in F_k \rangle$.

I. N. Sanov ([2] pp 324-325, or [3]) has shown that B(k) is finite. His proof gives very little further information on B(k). If G is any group let G_n be the nth term in its lower central series. Sanov's proof shows that since B(k) is a finite 2-group, and so nilpotent, then $B(k)_N = 1$ for some sufficiently large N. The most important result is due to C. R. B. Wright [5,6] who has shown that $B(k)_{3k} = 1$.

Let us define a sequence of formal words in symbols $a_1, a_2, \ldots, y$.

$$W_1(y) = y^4$$

$$W_2(a_1,y) = W_1(a_1)^{-1}W_1(y)^{-1}W_1(a_1y)$$

$$W_3(a_1,a_2,y) = W_2(a_1,a_2)^{-1}W_2(a_1,y)^{-1} W_2(a_1,a_2y)$$

$$W_r(a_1,a_2,\ldots,a_{r-1},y) =$$

$$W_{r-1}(a_1,a_2,\ldots,a_{r-2},a_{r-1})^{-1}W_{r-1}(a_1,a_2,\ldots,a_{r-2},y)^{-1}$$

$$W_{r-1}(a_1,a_2,\ldots,a_{r-2},a_{r-1}y).$$

Let G be a group generated by $a_1, a_2, \ldots, y$ which is therefore a factor group of the free group F on these generators. Using Philip Hall's collection process to express the W's in terms of commutators [2, pp 165-168] it follows that if W_i can be expressed in terms of commutators with leading terms commutators of weight $w(i)$, then $w(i) \geq i$, since replacing any one of $a_1,\ldots,a_{i-1},y$ by the identity makes W_i the identity in the free group on the rest. Further if in G, $W_1(x_1) = 1$ for x_1 any generator of G and if $W_2(x_1,y) = 1$ for x_1 a generator and y an arbitrary element of G, induction on the length of y shows that $W_1(y) = 1$ in G for any y in G. Similarly using induction on the length of y, if $W_{i-1}(x_1,\ldots,x_{i-1}) = 1$ for all choices of $x_1,\ldots,x_{i-1}$ as generators of and if also $W_i(x_1,\ldots,x_{i-1},y) = 1$ for $x_1,\ldots,x_{i-1}$ generators of G and y arbitrary then $W_{i-1}(x_1,\ldots,x_{i-2},y) = 1$ for $x_1,\ldots,x_{i-2}$ generators and y arbitrary, and so in terms of the W's it will follow that also $W_1(y) = 1$ for y arbitrary, provided that $W_j(x_1,\ldots,x_j) = 1$ for $x_1,\ldots,x_j$ generators, $j = 1,\ldots,i-1$ and $W_i(x_1,\ldots,x_{i-1},y) = 1$ if $x_1,\ldots,x_{i-1}$ are generators and y arbitrary.

These observations lead to a finite presentation of B(k) as stated in the following theorem:

Theorem 2.1. Let B(k) be the Burnside group of exponent 4 with generators $x_1, \ldots, x_k$. Then B(k) is defined by the following finite set of relations:

(i) $(a_1,\ldots,a_r)^4 = 1 \quad r = 1, \ldots, 3k-1$ and $a_1, \ldots, a_r$ generators $x_1, \ldots, x_k$ in some order

and (ii) $(a_1, a_2, \ldots, a_{3k}) = 1$ with $a_1, \ldots, a_{3k}$ generators $x_1, \ldots, x_k$ in some order.

Proof: Let G = B(k) the group of exponent 4 generated by $x_1, \ldots, x_k$. By Wright's result $G_{3k} = 1$ and for this the conditions (ii) are sufficient.

This means that $W_{3k}(a_1, a_2, \ldots, a_{3k-1}, y) = 1$ for $a_1, \ldots, a_{3k-1}$ generators and y arbitrary. From (i) we have $W_1(a_1) = 1$, $W_1(y) = 1$ and $W_1(a_1y) = 1$ so that $W_2(a_1, y_2) = 1$ with a_1 a generator and y a product of at most 3k - 2 generators. By induction we have $W_r(a_1, a_2, \ldots, a_{r-1}, y) =$ for $a_1, \ldots, a_{r-1}$ generators and y a product of at most 3k - r generators so that $W_{3k-1}(a_1, \ldots, a_{3k-1}) = 1$ for $a_1, \ldots, a_{3k-1}$ generators. But now since from (ii) $G_{3k} = 1$ we have $W_{3k}(a_1, \ldots, a_{3k-1}, y) = 1$ with $a_1, \ldots, a_{3k-1}$ generators and y arbitrary. But now as

$$W_{r-1}(a_1, a_2, \ldots, a_{r-2}, a_{r-1}y) = W_r(a_1, a_2, \ldots, a_{r-2}, y)W_{r-1}(a_1, a_2, \ldots, a_{r-1})$$

$W_r(a_1, a_2, \ldots, a_{r-1}, y)$ using induction on the length of y we may conclude from $W_{3k}(a_1, \ldots, a_{3k-1}, y) = 1$ for y arbitrary that also (taking r = 3k) $W_{3k-1}(a_1, a_2, \ldots, a_{3k-2}, y) = 1$ with $a_1, \ldots, a_{3k-2}$ generators and y arbitrary, and going from a value of r to a value one less that $W_r(a_1, \ldots, a_{r-1}, y) = 1$ for $a_1, \ldots, a_{r-1}$ generators and y arbitrary. Ultimately we conclude that $W_1(y) = y^4 = 1$ for y arbitrary so that these relations do determine $G = \langle x_1, \ldots, x_k \rangle$ as B(k).

The following lemma, due to S. Tobin [4], is useful when we wish to make substitutions in identities.

Lemma: Let $A_1A_2 \cdots A_m = B_1 \cdots B_s$ be an identity on commutators valid in any group G of exponent 4. Suppose that each of $x_1, \ldots, x_k$, a subset of the generators of G appears as a component in every commutator $A_1, \ldots, A_m$ and suppose that every commutator $B_1 \cdots B_s$ is of weight at least n. Then without loss of generality we may assume that every commutator $B_1, \ldots, B_s$ contains each of $x_1, \ldots, x_k$ as a component.

Proof: If certain of the B's, say $B_{i_1} B_{i_2} \cdots B_{i_r}$, do not contain x_1 as a component, then we may use the collection process to move these to the front putting the identity in the form $A_1 \cdots A_m = B_{i_1} B_{i_2} \cdots B_{i_r} C_1 \cdots C_s$ where every C contains x_1 as a component. Since this is an identity, we may put $x_1 = 1$ whence $A_1 = \cdots = A_m = 1$ and $C_1 \cdots C_s = 1$ giving $1 = B_{i_1} B_{i_2} \cdots B_{i_r}$ an identity which does not involve x_1 and so $A_1 \cdots A_m = C_1 \cdots C_s$ is an identity. Continuing we may reach an identity in which every commutator on the right contains every one of $x_1, \ldots, x_k$ as a component.

3. Some basic relations for groups of exponent 4

The group $G = \langle a,b \rangle$ with following defining relations is in fact $B(2)$. The relations are:

$$3.1) \qquad a^4 = b^4 = (ab)^4 = (a^{-1}b)^4 = (a^2b)^4 = (ab^2)^4 = 1$$
$$(a^{-1}b^{-1}ab)^4 = (a^{-1}bab)^4 = 1.$$

Using coset enumeration of cosets of the subgroup $H = \langle a^2,b \rangle$ we find that G has 64 cosets and H has order 64, so that G has order 2^{12}. Since it is not hard to show that H has order at most 64, this representation for G is faithful. We may use this representation to show that $G_6 = 1$ and that all elements of length at most 5 have order 2 or 4. Thus by Theorem 2.1, $G = B(2)$ and is of order 2^{12}. In his original paper Burnside [1] showed that $B(2)$ has order at most 2^{12}. In his Ph.D. thesis (Manchester University 1955) Sean Tobin proved that the order of $B(2)$ is exactly 2^{12} in another way.

In terms of commutators the relations for $B(2)$ are the following:

$$B(2) = \langle a,b\rangle$$

$$a^4 = 1 \ , \ b^4 = 1$$

$$(b,a)^2 = (b,a,a,a)(b,a,a,b)(b,a,b,b)(b,a,b;b,a)$$

$$(b,a,a)^2 = (b,a,a;b,a)$$

$$(b,a,b)^2 = (b,a,b;b,a)$$

$$(b,a,a,a)^2 = 1$$

$$(b,a,a,b)^2 = 1$$

3.2) $$(b,a,b,b)^2 = 1$$

$$(b,a,a,a,a) = (b,a,a;b,a)$$

$$(b,a,a,a,b) = (b,a,a;b,a)(b,a,b;b,a)$$

$$(b,a,a,b,b) = (b,a,a;b,a)$$

$$(b,a,b,b,b) = (b,a,b;b,a)$$

$$(b,a,a;b,a)^2 = 1$$

$$(b,a,b;b,a)^2 = 1$$

$$B(2)_6 = 1$$

All commutator relations in this paper can be derived from the following basic relations for groups of exponent four:

3.3.1) $(b,a)^2 = (b,a,a,a)(b,a,a,b)(b,a,b,b)(b,a,b;b,a)$

3.3.2) $(a,b,c)(b,c,a)(c,a,b) \equiv 1 \pmod{G_4}$

3.3.3) $(c,a,a,b)(b,a,a,c)(c,a,b,b)(b,a,c,c)(c,a;b,a)(c,b;b,a)(c,b;c,a) \equiv 1 \pmod{G_5}$

3.3.4) $(d,a;c,b)(d,b;c,a)(d,c;b,a) \equiv 1 \pmod{G_5}$

3.3.5) $(a,b;c,d;f)(a,d;c,f;b)(a,f;c,b;d) \equiv 1 \pmod{G_6}$

Here 3.3.1) is the first relation in 3.2), 3.3.2) is the Jacobi identity, valid in any group. Here 3.3.3), 3.3.4) and 3.3.5) are the first significant terms of $(abc)^4$, $(abcd)^4$, and $(abcdf)^4$ as collected by Philip Hall's process. We shall refer to 3.3.5) as Wright's relations, since he was the first to find and exploit it. For some reason the relation 3.3.3) does not appear in his paper.

In addition to the above, there are of course the universally valid identities

3.4.1) $(y,x) = (x,y)^{-1}$,

3.4.2) $(xy,z) = (x,z)(x,z,y)(y,z)$,

3.4.3) $(x,yz) = (x,z)(x,y)(x,y,z)$.

A consequence of 3.3.1) is of course that $(x_1,x_2,\ldots,x_n)^2 \equiv 1 \pmod{G_{n+2}}$. From this, 3.4.1) and 3.3.2) we have $(x;y,z) \equiv (y,z,x)^{-1} \equiv (y,z,x) \equiv (z,x,y)(x,y,z) \equiv (x,z,y)(x,y,z) \pmod{G_4}$. The rule

3.5) $(x;y,z) \equiv (x,y,z)(x,z,y) \pmod{G_4}$

enables us to express complex commutators in terms of simple commutators. Also 3.5) enables us to express any commutator which involves a particular generator, say c, in terms of commutators with c as the first entry. In this way we may rewrite 3.3.3) in the form

3.6) $(c,a,a,b)(c,a,b,a)(c,b,a,a)(c,a,b,b)(c,b,a,b)(c,b,b,a)(c,a,b,c)(c,b,c,a)$
$\equiv 1 \pmod{G_5}$.

The following formulas are (6), (7) and (9) from Wright's paper.

3.7) $(x_1,\ldots,x_i,a,a,x_{i+1},\ldots,x_n) \equiv (x_1,\ldots,x_i,a^2,x_{i+1}, x_n)(\text{mod } G_{n+3})$.

3.8) $(x,y,a,a,a,z) \equiv (x,y,z,a,a,a)(\text{mod } G_7)$.

3.9) $(x,y,a,a,z,a,w) \equiv (x,y,z,a,a,w,a)(\text{mod } G_8)$.

In 3.3.5) if we put $a = x$, $b = y$, $c = y$, $d = y$ and $f = z$ we have

3.10) $(x,y,y,z,y)(x,y,z,y,y) \equiv 1 \ (\text{mod } G_6)$.

Wright's relation (8) is obtainable from this replacing x by (x,y) and y by a.

In 3.6) let us replace c by (c,b) giving

3.11) $(c,b,a,a,b)(c,b,a,b,a)(c,b,b,a,a)(c,b,a,b,b)(c,b,b,a,b)(c,b,b,b,a)$
$\equiv 1 \ (\text{mod } G_6)$.

From 3.10) $(c,b,a,b,b)(c,b,b,a,b) \equiv 1 \ (\text{mod } G_6)$ so that 3.11) simplifies to

3.12) $(c,b,a,a,b)(c,b,a,b,a)(c,b,b,a,a)(c,b,b,b,a) \equiv 1 \ (\text{mod } G_6)$.

In this replace a by a^2 whence from 3.7)

3.13) $(c,b,b,b,a,a) \equiv 1 \ (\text{mod } G_7)$.

Take the commutator of 3.12) with a giving

3.14) $(c,b,a,a,b,a)(c,b,a,b,a,a)(c,b,b,a,a,a)(c,b,b,b,a,a) \equiv 1 \ (\text{mod } G_7)$.

From 3.10) the product of the first two of these is the identity and using 3.13) this reduces to

3.15) $(c,b,b,a,a,a) \equiv 1 \ (\text{mod } G_7)$.

In this put $c = z$ and $b = xy$. Expanding gives

3.16) $(z,x,x,a,a,a)(z,x,y,a,a,a)(z,y,x,a,a,a)(z,y,y,a,a,a) \equiv 1 \pmod{G_7}$.

Here the first and fourth terms are 1 by 3.15) whence

3.17) $(z,x,y,a,a,a)(z,y,x,a,a,a) \equiv 1 \pmod{G_7}$.

But from 3.5) $(z,y,x)(z,x,y) \equiv (z;x,y) \equiv (x,y,z) \pmod{G_4}$ and so 3.17) becomes

3.18) $(x,y,z,a,a,a) \equiv 1 \pmod{G_7}$.

From 3.8) we then also have

3.19) $(x,y,a,a,a,z) \equiv 1 \pmod{G_7}$.

4. Commutator relations in B(3)

Using the relations of section 3 we can find a commutator basis for elements of G = B(3) which we suppose generated by elements a,b,c. For weights 1, 2, and 3 a basis is given by

a,	(b,a,a),
b,	(b,a,b),
c,	(c,a,a),
(b,a),	(c,a,b),
(c,a),	(c,a,c),
(c,b),	(c,b,a),
	(c,b,b),
	(c,b,c).

Here a, b, and c are order 4 modulo $G' = G_2$, but any commutator of weight $n \geq 2$ is of order 2 modulo G_{n+1}, by 3.3.1). There are no other relations on commutators of weight at most 3 so that G/G_4 is of order 2^{17}.

Commutators of weight 4 modulo G_5 are given in the following

(b,a,a,a),
(b,a,a,b)
(b,a,b,a) ≡ (b,a,a,b)

4.2)

(c,a,a,a)	(c,b,a,a)
(c,a,a,b)	(c,b,a,b)
(c,a,a,c)	(c,b,a,c)
(c,a,b,a)	(c,b,b,a)
(c,a,b,b)	(c,b,b,b)
(c,a,b,c)	(c,b,b,c)
(c,a,c,a) ≡ (c,a,a,c)	(c,b,c,a)
(c,a,c,b)	(c,b,c,b) ≡ (c,b,b,c)
(c,a,c,c)	(c,b,c,c)

These also satisfy relations modulo G_5.

4.3) $$(c,a,b,c)(c,a,c,b)(c,b,a,c)(c,b,c,a) \equiv 1,$$

$$(c,a,a,b)(c,a,b,a)(c,b,a,a)(c,a,b,b)(c,b,a,b)(c,b,b,a)(c,a,b,c)(c,b,c,a) \equiv 1.$$

The Jacobi relation $(x,y,z)(y,z,x)(z,x,y) \equiv 1$ with $x = (b,a)$, $y = b$, $z = a$ gives

$$(b,a,b,a)(b,a;b,a)(a;b,a;b) \equiv 1 \quad \text{or}$$

$(b,a,b,a)(b,a,a,b) \equiv 1$ accounting for the relations in 4.2).

The relation $(c,a;c,b)(c,b;c,a) \equiv 1$ when expanded gives the first relation in 4.3) and the second is the relation 3.6). Thus the order of G_4 modulo G_5 is at most 2^{17}.

Calculation of G_5 modulo G_6 is much more complicated. There is exactly 6 independent involving only two of the generators, as may be seen from the relations for B(2). These are

(4.4)
$$\begin{aligned}
(b,a,a,a,a) &= (b,a,a;b,a)\\
(b,a,b,b,b) &= (b,a,b;b,a)\\
(c,a,a,a,a) &= (c,a,a;c,a)\\
(c,a,c,c,c) &= (c,a,c;c,a)\\
(c,b,b,b,b) &= (c,b,b;c,b)\\
(c,b,c,c,c) &= (c,b,c;c,b)
\end{aligned}$$

Simple commutators of weight 5 involving all three generators are given in the following table.

		Interchange (a,b)	Interchange (a,c)
	$r_1 = (c,a,a,a,b)$	r_{31}	r_{18}
	$r_2 = (c,a,a,b,a)$	r_{29}	r_{17}
	$r_3 = (c,a,a,b,b)$	r_{28}	r_{16}
	$r_4 = (c,a,a,b,c)$	r_{30}	r_{15}
	$r_5 = (c,a,a,c,b)$	r_{32}	r_5
	$r_6 = (c,a,b,a,a)$	r_{23}	r_{14}
	$r_7 = (c,a,b,a,b)$	r_{22}	r_{13}
	$r_8 = (c,a,b,a,c)$	r_{24}	r_{12}
4.5)	$r_9 = (c,a,b,b,a)$	r_{20}	r_{11}
	$r_{10} = (c,a,b,b,b)$	r_{19}	r_{10}
	$r_{11} = (c,a,b,b,c)$	r_{21}	r_9
	$r_{12} = (c,a,b,c,a)$	r_{26}	r_8
	$r_{13} = (c,a,b,c,b)$	r_{25}	r_7
	$r_{14} = (c,a,b,c,c)$	r_{27}	r_6
	$(c,a,c,a,b) \equiv (c,a,a,c,b) = r_5$		
	$r_{15} = (c,a,c,b,a)$	r_{34}	r_4
	$r_{16} = (c,a,c,b,b)$	r_{33}	r_3
	$r_{17} = (c,a,c,c,b)$	r_{35}	r_2
	$r_{18} , (c,a,c,c,b)$	r_{36}	r_1

		Interchange (a,b)	Interchange (a,c)
	r_{19} = (c,b,a,a,a)	r_{10}	$r_{14}r_{27}$
	r_{20} = (c,b,a,a,b)	r_{9}	$r_{13}r_{26}$
	r_{21} = (c,b,a,a,c)	r_{11}	$r_{12}r_{25}$
	r_{22} = (c,b,a,b,a)	r_{7}	$r_{11}r_{24}$
	r_{23} =)c,b,a,b,b)	r_{6}	$r_{10}r_{23}$
	r_{24} = (c,b,a,b,c)	r_{8}	$r_{9}\ r_{22}$
	r_{25} = (c,b,a,c,a)	r_{13}	$r_{8}\ r_{21}$
	r_{26} = (c,b,a,c,b)	r_{12}	$r_{7}\ r_{20}$
	r_{27} = (c,b,a,c,c)	r_{14}	$r_{6}\ r_{19}$
4.5)	r_{28} = (c,b,b,a,a)	r_{3}	$r_{11}r_{30}$
	r_{29} = (c,b,b,a,b)	r_{2}	$r_{10}r_{29}$
	r_{30} = (c,b,b,a,c)	r_{4}	$r_{9}\ r_{28}$
	r_{31} = (c,b,b,b,a)	r_{1}	$r_{10}r_{23}r_{29}r_{31}$
	r_{32} = (c,b,b,c,a)	r_{5}	$r_{3}\ r_{28}$
	r_{33} = (c,b,c,a,a)	r_{16}	$r_{4}\ r_{21}$
	r_{34} = (c,b,c,a,b)	r_{15}	$r_{3}\ r_{20}$
	r_{35} = (c,b,c,a,c)	r_{17}	$2_{2}\ r_{19}$
	(c,b,c,b,a) ≡ (c,b,b,c,a) = r_{32}		
	r_{36} = (c,b,c,c,a)	r_{18}	$r_{1}\ r_{2}\ r_{6}\ r_{19}$

Application of the Jacobi identity gives the following relations on the r's modulo G_6

$$r_{14}r_{17}r_{27}r_{35} \equiv 1$$
$$r_{17}r_{18}r_{27}r_{36} \equiv 1$$
$$r_{12}r_{15}r_{25}r_{33} \equiv 1$$
4.6)
$$r_4\ r_5\ r_{21}r_{31} \equiv 1$$
$$r_{11}r_{16}r_{30}r_{32} \equiv 1$$
$$r_{13}r_{16}r_{26}r_{34} \equiv 1$$

In the Wright identity 3.3.5) we replace a by c, c by a and b,d,f respectively by b,b,b;a,a,b;a,b,b;c,b,b;c,a,b,c,b,a and c,c,b. Then apply interchanges of a and b and a and c. These operations give the following relations on the r's modulo G_6

$$r_{23}r_{29} \equiv 1$$
$$r_2\ r_6 \equiv 1$$
$$r_3\ r_7\ r_{22}r_{28} \equiv 1$$
$$r_{24}r_{26}r_{30}r_{34} \equiv 1$$
4.7)
$$r_4\ r_8\ r_{26}r_{33} \equiv 1$$
$$r_{27}r_{35} \equiv 1$$
$$r_{14}r_{17} \equiv 1$$
$$r_4\ r_8\ r_{12}r_{15} \equiv 1$$
$$r_{13}r_{16}r_{24}r_{30} \equiv 1$$

In 3.6) we replace in turn a by a^2, b by b^2, and c by c^2. Then we replace c first by (c,a) and then by (c,b), and interchange a with b and also a with c to obtain further relations. Finally we take the commutator of 3.6) respectively with a, b, and c. These operations yield the following relations on the r's modulo G_6

$$
\begin{aligned}
& r_3 r_4 r_{20} r_{28} r_{33} \equiv 1 \\
& r_5 r_7 r_{20} r_{21} r_{22} \equiv 1 \\
& r_5 r_{15} r_{16} r_{32} r_{33} r_{34} \equiv 1 \\
& r_1 r_2 r_3 r_6 r_7 r_9 \equiv 1 \\
& r_{20} r_{22} r_{23} r_{28} r_{29} r_{31} \equiv 1 \\
4.8) \qquad & r_{11} r_{13} r_{16} r_{18} \equiv 1 \\
& r_{10} r_{16} r_{26} r_{31} \equiv 1 \\
& r_4 r_8 r_{21} r_{36} \equiv 1 \\
& r_1 r_{12} r_{19} r_{33} \equiv 1 \\
& r_2 r_6 r_9 r_{12} r_{19} r_{22} r_{28} r_{33} \equiv 1 \\
& r_3 r_7 r_{10} r_{13} r_{20} r_{23} r_{29} r_{34} \equiv 1 \\
& r_4 r_8 r_{11} r_{14} r_{21} r_{24} r_{30} r_{35} \equiv 1
\end{aligned}
$$

Combining the relations 4.6), 4.7) and 4.8) we find relations expressing all the r's in terms of 15 of them: Here an r_j is expressed in terms of certain r_i's with $i < j$.

$$
\begin{aligned}
r_6 &\equiv r_2\\
r_9 &\equiv r_1 r_3 r_7\\
r_{15} &\equiv r_4 r_8 r_{12}\\
r_{16} &\equiv r_1 r_5 r_8 r_{10} r_{11} r_{12} r_{13}\\
r_{17} &\equiv r_{14}\\
r_{18} &\equiv r_{11} r_{13} r_{16}\\
r_{21} &\equiv r_1 r_4 r_5 r_{12} r_{19}\\
r_{22} &\equiv r_1 r_4 r_7 r_{12} r_{19} r_{20}\\
r_{24} &\equiv r_4 r_7 r_{12} r_{13} r_{19} r_{20}\\
r_{25} &\equiv r_1 r_4 r_8 r_{12} r_{19}\\
r_{26} &\equiv r_3 r_7 r_{10} r_{16} r_{20}\\
r_{27} &\equiv r_1 r_5 r_8 r_{11} r_{12} r_{13} r_{14} r_{16} r_{19} \qquad (4.9)\\
r_{28} &\equiv r_3 r_7 r_{22}\\
r_{29} &\equiv r_{23}\\
r_{30} &\equiv r_{13} r_{16} r_{24}\\
r_{31} &\equiv r_3 r_7 r_{20}\\
r_{32} &\equiv r_{11} r_{13} r_{24}\\
r_{33} &\equiv r_1 r_{12} r_{19}\\
r_{34} &\equiv r_{13} r_{16} r_{26}\\
r_{35} &\equiv r_{27}\\
r_{36} &\equiv r_4 r_8 r_{21}
\end{aligned}
$$

Thus G_5 modulo G_6 has a basis of at most the 6 commutators of 4.4) and the 15 r's with i = 1,2,3,4,5,7,8,10,11,12,13,14,19,20,23. Hence G_5 modulo G_6 is of order at most 2^{21}.

In finding further relations on commutators in $G = B(3)$ we shall not rely on Wright's main theorem but only on certain specific relation, particularly those listed in Section 3.

Wright [6, p. 362] shows that a commutator of any one of the six commutators in 4.4) with a, b, or c is a congruent to the identity modulo G_7. Thus in examining G_6 modulo G_7 we need only consider commutators (r_i,a), (r_i,b) and (r_i,c) where r_i is one of the r's.

First we examine the 32 simple commutators of weight 6 containing exactly one c, which bay be taken in the form (c,x_1,x_2,x_3,x_4,x_5) with x_i = a or b. A number of these are congruent to the identity modulo G_7 as a consequence of the relations 3.13), 3.18) and 3.19). From 3.10) we have

4.10) $\quad (c,a,b,a,a) \equiv (c,a,a,b,a) \pmod{G_6}$.

Commuting this with a gives

4.11) $\quad (c,a,b,a,a,a) \equiv (c,a,a,b,a,a) \pmod{G_7}$.

By 3.18) the left hand side is congruent to the identity modulo G_7 and so also the right hand must be. In 4.10) if we replace c by (c,a) we obtain

4.12) $\quad 1 \equiv (c,a,a,b,a,a) \equiv (c,a,a,a,b,a) \pmod{G_7}$.

We take the weight 5 relation

4.13) $\quad (c,a,a,c,b)(c,a,b,a,b)(c,b,a,a,b)(c,b,a,a,c)(c,b,a,b,a) \equiv 1 \pmod{G_6}$

and replace c by (c,a) to obtain

4.14) $(c,a,a,b,a,b)(c,a,b,a,a,b)(c,a,b,a,b,a) \equiv 1 \pmod{G_7}$.

Since by 3.10) the first two terms are equal we find

4.15) $(c,a,b,a,b,a) \equiv 1 \pmod{G_7}$.

In 3.13) replace b by ab and expand to get

4.16) $$(c,a,a,a,a,a)(c,a,a,b,a,a)(c,a,b,a,a,a)(c,a,b,b,a,a)$$
$$(c,b,a,a,a,a)(c,b,a,b,a,a)(c,b,b,a,a,a)(c,b,b,b,a,a) \equiv 1.$$

We know that all but two of these are the identity and so we have

4.17) $(c,a,b,b,a,a)(c,b,a,b,a,a) \equiv 1 \pmod{G_7}$.

Similarly in $(c,a,a,b,b,b) \equiv 1$ we replace b by ab and expand to obtain

4.18) $(c,a,a,b,a,b)(c,a,a,b,b,a) \equiv 1 \pmod{G_7}$

and also doing the same with $(c,a,b,b,b,a) \equiv 1$ gives

4.19) $(c,a,a,b,b,a)(c,a,b,b,a,a) \equiv 1 \pmod{G_7}$.

We may interchange a and b in the above relations and also make use of 3.10). This procedure gives exactly two different values for the commutators with exactly one c. These we call t_1 and t_2 and each is represented in six ways:

4.20)
$$\begin{aligned}
t_1 &\equiv (c,a,a,b,a,b), & t_2 &\equiv (c,a,b,a,b,b)\\
t_1 &\equiv (c,a,a,b,b,a), & t_2 &\equiv (c,a,b,b,a,b)\\
t_1 &\equiv (c,a,b,a,a,b), & t_2 &\equiv (c,b,a,a,b,b)\\
t_1 &\equiv (c,a,b,b,a,a), & t_2 &\equiv (c,b,a,b,b,a)\\
t_1 &\equiv (c,b,a,a,b,a), & t_2 &\equiv (c,b,b,a,a,b)\\
t_1 &\equiv (c,b,a,b,a,a), & t_2 &\equiv (c,b,b,a,b,a)
\end{aligned}$$

The remaining 20 commutators (c,x_1,x_2,x_3,x_4,x_5) with x_i = a or b are all found to be congruent to the identity modulo G_7 by the arguments above.

Commutators with exactly one a go into commutators with exactly one c on interchanging a and c. However with c as the first component there are Jacobi identities on the different expressions. With exactly one a there are two different values which we write t_3 and t_4 given in the following list

4.21)

$$\begin{array}{ll}
t_3 \equiv (c,a,b,b,c,b) & t_4 \equiv (c,a,b,b,c,c) \\
t_3 \equiv (c,a,b,c,b,b) & t_4 \equiv (c,a,b,c,c,b) \\
t_3 \equiv (c,b,a,b,b,c) & t_4 \equiv (c,a,c,b,b,c) \\
t_3 \equiv (c,b,a,b,c,b) & t_4 \equiv (c,a,c,b,c,b) \\
t_3 \equiv (c,b,b,a,b,c) & t_4 \equiv (c,b,a,c,b,c) \\
t_3 \equiv (c,b,b,c,a,b) & t_4 \equiv (c,b,a,c,c,b) \\
t_3 \equiv (c,b,c,a,b,b) & t_4 \equiv (c,b,b,c,a,c) \\
t_3 \equiv (c,b,c,b,a,b) & t_4 \equiv (c,b,c,a,c,b) \\
 & t_4 \equiv (c,b,c,b,a,c) \\
 & t_4 \equiv (c,b,b,a,c,c)
\end{array}$$

Other commutators involving exactly one a are congruent to the identity modulo G_7. Interchanging a and b we find the commutators with exactly one b. These are

$$t_5 \equiv (c,a,a,b,a,c) \qquad t_6 \equiv (c,a,a,b,c,c)$$

$$t_5 \equiv (c,a,a,c,b,a) \qquad t_6 \equiv (c,a,a,c,b,c)$$

$$t_5 \equiv (c,a,b,a,a,c) \qquad t_6 \equiv (c,a,b,c,a,c)$$

$$t_5 \equiv (c,a,b,a,c,a) \qquad t_6 \equiv (c,a,b,c,c,a)$$

$$t_5 \equiv (c,a,c,a,b,a) \qquad t_6 \equiv (c,a,c,a,b,c)$$

$$t_5 \equiv (c,a,c,b,a,a) \qquad t_6 \equiv (c,a,c,b,c,a)$$

$$t_5 \equiv (c,b,a,a,c,a) \qquad t_6 \equiv (c,b,a,a,c,c)$$

$$t_5 \equiv (c,b,a,c,a,a) \qquad t_6 \equiv (c,b,a,c,c,a)$$

$$t_6 \equiv (c,b,c,a,a,c)$$

$$t_6 \equiv (c,b,c,a,c,a)$$

It remains to consider commutators of weight six each containing two a's, two b's, and two c's. These are

	Interchange (a,b)	Interchange (a,c)
$s_1 = (c,a,a,b,b,c)$	s_{19}	s_{12}
$s_2 = (c,a,a,b,c,b)$	s_{20}	s_{11}
$s_3 = (c,a,a,c,b,b)$	s_{21}	s_3
$s_4 = (c,a,b,a,b,c)$	s_{15}	s_9
$s_5 = (c,a,b,a,c,b)$	s_{16}	s_8
$s_6 = (c,a,b,b,a,c)$	s_{13}	s_7
$s_7 = (c,a,b,b,c,a)$	s_{14}	s_6
$s_8 = (c,a,b,c,a,b)$	s_{18}	s_5
$s_9 = (c,a,b,c,b,a)$	s_{17}	s_4
$s_{10} = (c,a,c,a,b,b) \equiv (c,a,a,c,b,b) = s_3$		
$s_{11} = (c,a,c,b,a,b)$	s_{23}	s_2
$s_{12} = (c,a,c,b,b,a)$	s_{22}	s_1

4.23)

$s_{13} = (c,b,a,a,b,c)$	s_6	$s_9 s_{18}$
$s_{14} = (c,b,a,a,c,b)$	s_7	$s_8 s_{17}$
$s_{15} = (c,b,a,b,a,c)$	s_4	$s_7 s_{16}$
$s_{16} = (c,b,a,b,c,a)$	s_5	$s_6 s_{15}$
$s_{17} = (c,b,a,c,a,b)$	s_9	$s_5 s_{14}$
$s_{18} = (c,b,a,c,b,a)$	s_8	$s_4 s_{13}$
$s_{19} = (c,b,b,a,a,c)$	s_1	$s_7 s_{20}$
$s_{20} = (c,b,b,a,c,a)$	s_2	$s_6 s_{19}$
$s_{21} = (c,b,b,c,a,a)$	s_3	$s_1 s_{19}$
$s_{22} = (c,b,c,a,a,b)$	s_{12}	$s_2 s_{14}$
$s_{23} = (c,b,c,a,b,a)$	s_{11}	$s_1 s_{13}$
$s_{24} = (c,b,c,b,a,a) \equiv (c,b,b,c,a,a) = s_{21}$		

The Jacobi identity yields the following relations on the s's modulo G_7.

4.24)
$$s_8 s_{11} s_{17} s_{22} \equiv 1$$
$$s_2 s_3 \; s_{14} s_{22} \equiv 1$$
$$s_9 s_{12} s_{18} s_{23} \equiv 1$$
$$s_7 s_{12} s_{20} s_{21} \equiv 1$$

In 3.3.3) replace a by a^2 to obtain

4.25) $(c,a,a,b,b)(c,b,a,a,b)(c,b,b,a,a)(c,a,a,b,c)(c,b,c,a,a) \equiv 1 (\text{mod } G_6)$.

In this replace b by b^2 to obtain modulo G_7

4.26) $(c,a,a,b,b,c)(c,b,b,c,a,a) \equiv 1$, or $s_1 s_{21} \equiv 1$.

Taking the commutator of 3.25) with b gives

4.27) $(c,a,a,b,b,b)(c,b,a,a,b,b)(c,b,b,a,a,b)(c,a,a,b,c,b)(c,b,c,a,ab) \equiv 1$,

or $1 \cdot t_2 \cdot t_2 \cdot s_2 \cdot s_{23} \equiv 1$, or $s_2 s_{22} \equiv 1$.

Similarly the commutators of 4.26) with c gives

4.28) $s_1 s_{13} s_{19} \equiv 1$.

In Wright's relation 3.3.5) replace a by (c,a), b by a, c by b, d by c and f by b. This gives

4.29) $s_2 s_3 s_4 s_6 \equiv 1$.

With these relations and others obtained by repeated interchanges of a and b or of a and c the following relations may be found modulo G_7.

$$s_1 s_2 s_3 \equiv 1,$$

4.30) $$s_1 \equiv s_5 \equiv s_7 \equiv s_{11} \equiv s_{15} \equiv s_{17} \equiv s_{21} \equiv s_{24},$$
$$s_2 \equiv s_6 \equiv s_8 \equiv s_{12} \equiv s_{13} \equiv s_{18} \equiv s_{20} \equiv s_{22},$$
$$s_3 \equiv s_4 \equiv s_9 \equiv s_{10} \equiv s_{14} \equiv s_{16} \equiv s_{19} \equiv s_{23}.$$

Hence G_6 modulo G_7 has order at most 2^8 with a basis $t_1,\ldots,t_6, s_1, s_2$.

The effect of interchanges on the t's and s's is given by the following table.

4.31)

	Interchange (a,b)	Interchange (a,c)
t_1	t_2	t_4
t_2	t_1	t_3
t_3	t_5	t_2
t_4	t_6	t_1
t_5	t_3	t_6
t_6	t_4	t_5
s_1	s_3	s_2
s_2	s_2	s_1
s_3	s_1	s_3

The commutators of G_7 modulo G_8 are commutators of the t's and s's with a, b, and c.

We now find for t_1, using the most convenient form

4.32) $$(t_1,a) \equiv (c,a,b,b,a,a,a) \equiv 1,$$
$$(t_1,b) \equiv (c,a,a,b,b,a,b) \equiv (c,a,b,b,a,b,a) \equiv (t_2,a),$$
$$(t_1,c) \equiv (c,b,a,a,b,a,c) \equiv (c,b,b,a,a,c,a) \equiv (s_{19},a) \equiv (s_3,a).$$

We also find

4.33) $$(s_1,a) \equiv (s_{21},a) \equiv (c,b,b,c,a,a,a) \equiv 1.$$

Here we have used the shifting relation 3.9) and the Engel type relation 3.18).

Applying interchanges of a and b and of a and c to the above relations we find a basis of at most 6 commutators in a basis for G_7 modulo G_8. These are given here, noting that from $(s_1,a) \equiv 1$ and $s_1s_2s_3 \equiv 1$ we have $(s_2,a) \equiv (s_3,a)$.

4.34)

	Int(a,b)	Int(a,c)
$q_1 \equiv (s_2,a) \equiv (s_3,a) \equiv (t_1,c) \equiv (t_5,b)$	q_2	q_3
$q_2 = (s_1,b) \equiv (s_2,b) \equiv (t_2,c) \equiv (t_3,a)$	q_1	q_2
$q_3 \equiv (s_1,c) \equiv (s_3,c) = (t_4,a) \equiv (t_6,b)$	q_3	q_1
$q_4 \equiv (t_1,b) \equiv (t_2,a)$	q_4	q_5
$q_5 \equiv (t_3,c) \equiv (t_4,b)$	q_6	q_4
$q_6 \equiv (t_5,c) \equiv (t_6,a)$	q_5	q_6 .

To examine G_8 modulo G_9 we find

4.35) $$(q_4,x) \equiv (c,a,a,b,b,a,b,x) \equiv (c,a,a,a,b,b,x,b) \equiv (\text{mod } G_3)$$

from the shifting relation 3.9) and the relation 3.13) with a and b interchanged. Applying interchanges this gives modulo G_9.

4.36)
$$(q_4,a) \equiv (q_4,b) \equiv (q_4,c) \equiv 1$$
$$(q_5,a) \equiv (q_5,b) \equiv (q_5,c) \equiv 1$$
$$(q_6,a) \equiv (q_6,b) \equiv (q_6,c) \equiv 1 .$$

We now calculate

$$
\begin{aligned}
&(q_1,a) \equiv (s_2,a,a) \equiv (s_{12},a,a) \equiv (c,a,c,b,b,a,a,a) \equiv 1\\
&(q_1,b) \equiv (s_2,a,b) \equiv (s_{22},a,b) \equiv (c,b,c,a,a,b,a,b)\\
4.37)\quad &\qquad \equiv (c,b,c,b,a,a,b,a) \equiv s_{24},b,a) \equiv s_1,b,a) \equiv (q_2,a)\\
&(q_1,c) \equiv (s_3,a,c) \equiv (s_{19},a,c) \equiv (c,b,b,a,a,c,a,c)\\
&\qquad \equiv (c,b,b,c,a,a,c,a) \equiv (s_{21},c,a) \equiv (s_1,c,a) \equiv (q_3,a).
\end{aligned}
$$

Applying interchanges we find that G_8 modulo G_9 has a basic of at most 3 elements and order at most 2^3

$$
\begin{aligned}
&w_1 \equiv (q_2,c) \equiv (q_3,b)\\
4.38)\quad &w_2 \equiv (q_1,c) \equiv (q_3,a)\\
&w_3 \equiv (q_1,b) \equiv (q_2,a) \quad .
\end{aligned}
$$

Finally we calculate G_9 modulo G_{10}. We have

4.39) $\quad (w_3,a) \equiv (q_2,a,a) \equiv (c,b,c,b,a,a,b,a,a) \equiv (c,b,c,b,a,b,a,a,a) \equiv$

using 3.10) and 3.18). Also

4.40) $\quad (w_3,b) \equiv (q_1,b,b) \equiv (t_5,b,b,b) \equiv 1 \ (G_{10}).$

Then

$$
\begin{aligned}
4.41)\quad (w_3,c) &\equiv (q_1,b,c) \equiv t_5,b,b,c) \equiv (c,a,a,c,b,a,b,b,c)\\
&\equiv (c,a,a,c,a,b,c,b,b) \equiv 1 \ (G_{10})
\end{aligned}
$$

using the shifting relation and Wright's relation (p.392) that $(c,a,a,c,a,b) \equiv 1 \ (G_7)$.

Applying interchanges we find that every element of G_9 is congruent to 1 modulo G_{10}, and so from Sanov's result $G_9 = B(3)_9 = 1$.

Thus we have found bounds on G_i mod g_{i+1} and $G_9 = 1$. The bounds are

	G_1 mod G_2	2^6
	G_2 mod G_3	2^3
	G_3 mod G_4	2^8
	G_4 mod G_5	2^{17}
4.42)	G_5 mod G_6	2^{21}
	G_6 mod G_7	2^8
	G_7 mod G_8	2^6
	G_8 mod G_9	2^3
	$G_9 = 1$	

From this it follows that the order of G divides 2^{72}.

References

1. W. Burnside, "On an unsettled question in the theory of discontinuous groups". Quart. J. Pure and Appl. math. 33 (1902), 230-238.

2. M. Hall, Jr., "The Theory of Groups", The Macmillan Co., 1959.

3. I. N. Sanov, "Solution of Burnside's problem for exponent 4", Leningrad State Univ. Ann. 10 (1940), 166-170.

4. S. Tobin, "On a theorem of Baer and Higman", Canadian J. Math. 8(1956), 263-270.

5. C. R. B. Wright, "On groups of exponent four with generators of order two", Pacific J. Math. 10 (1960), 1097-1105.

6. ____________, "On the nilpotency class of a group of exponent four", Pacific J. Math. 11 (1961), 387-394.

PRIMARY ABELIAN GROUPS AND THE NORMAL STRUCTURE OF THEIR AUTOMORPHISM GROUPS

Jutta Hausen[1]

The present discussion is concerned with the normal structure of the group $\Gamma = \text{Aut } A$ of all automorphisms of a reduced abelian p-group A.

Particular attention is given to the question of how the primary normal subgroups of Γ are situated within the lattice of all normal subgroups of Γ. In [4] we proved the following result.

THEOREM 0 ([4], p. 276). If $p > 3$ then every normal torsion subgroup of Γ without elements of order p is contained in the center of Γ.

This theorem permits us to restrict our attention to the sublattice of all normal p-subgroups of Γ.

If A is elementary abelian then the lattice of all normal subgroups of Γ is completely known (see [2]; [7]). In this case, A is a vector-space over the prime field of characteristic p and $\Gamma = GL(A)$ is the group of all invertible linear transformations of A. In particular, Γ contains no normal p-subgroup $\neq 1$.

For $pA \neq 0$ however, Γ contains a large amount of normal p-subgroups. If B and C are characteristic subgroups of A and $B \leq C$, then the set

---

1) The author acknowledges support from the National Science Foundation Grant GP-34195.

of all $\gamma \in \Gamma$ such that γ induces the 1-automorphism in C/B is a normal subgroup of Γ which shall be denoted by $\sum(C/B)$. Let $\sum(A:C) = \sum(A/C) \cap \sum(C)$. It is well known that

$$\sum(A:C) \cong \mathrm{Hom}(A/C,C)$$

(see [3], p. 153). Therefore, every characteristic subgroup C of A gives rise to a (commutative) normal p-subgroup of Γ, namely the maximal torsion subgroup of $\sum(A:C)$. In particular, if A/C or C is bounded, then $\sum(A:C)$ is a p-group.

Our main result is the following theorem. We call a subgroup S of a group X noncentral if S is not contained in the center of X.

THEOREM 1. If $p > 3$ and $pA \neq 0$ then every noncentral normal subgroup of Γ contains a noncentral normal p-subgroup of Γ.

For $pA = 0$ and $p > 3$, every noncentral normal subgroup of $\Gamma = GL(A)$ contains the group $SL(A)$ of all linear transformations of A with determinant 1, which is generated by elements of order p (see [2], pp. 41,45). This property of general linear groups together with Theorem 1 leads to the following result which represents a considerable improvement of Theorem 0.

COROLLARY 2. If $p > 3$ then every normal subgroup of Γ without elements of order p is contained in the center of Γ.

The center, $\underline{z}\Gamma$, of Γ has been determined in [1]. It consists precisely of the multiplications with units in the ring of p-adic integers. In particular, $\gamma \in \Gamma$ belongs to the center of Γ if (and only if) γ induces the identity mapping in the lattice of all subgroups of A ([1], pp. 110,111).

It follows that Γ is commutative (if and) only if A is cyclic.

Using this result together with Theorem 1 and the properties of general linear groups mentioned above, the equivalence of conditions (i)-(iv) in the following theorem becomes obvious. Condition (v) involves the maximal normal p-subgroup, $P\Gamma$, of Γ. We note that in [5] we described $P\Gamma$ by means of its action on A.

THEOREM 3. If $p > 3$ and A is not cyclic, then the following conditions are equivalent.

(i) $pA \neq 0$.

(ii) Every noncentral normal subgroup of Γ contains a noncentral normal p-subgroup of Γ.

(iii) Γ contains a noncentral normal p-subgroup.

(iv) Γ contains a normal p-subgroup $\neq 1$.

(v) The centralizer, $\underline{c}P\Gamma$, of $P\Gamma$ in Γ is abelian.

The implication (v) $\to$ (i) in Theorem 3 is easily established, even without any restriction on the prime p: if $pA = 0$, then $P\Gamma = 1$ and

$$\underline{c}P\Gamma = \Gamma = GL(A),$$

which is abelian only if A is cyclic. The reverse implication follows from the fact that, if $pA \neq 0$, then

$$\underline{c}P\Gamma \leq \underline{z}\Gamma \cdot \sum(A:A[p]).$$

We point out that Theorems 1 and 3 are only corollaries to more general results on the automorphism group of an arbitrary, not necessarily reduced, abelian p-group. The details of these results will appear elsewhere ([6]).

REFERENCES

[1] R. Baer, Primary abelian groups and their automorphisms, Amer. J. Math., 59 (1937), 99-117.

[2] J. Diendonné, Les determinants sur un corps non commutatif, Bull. Soc. Math., France 71 (1943), 27-45.

[3] J. Hausen, Automorphismengesättigte Klassen abzählbarer abelscher Gruppen, Studies on Abelian Groups, New York 1968, pp. 147-181.

[4] ________, Abelian torsion groups with artinian primary components and their automorphisms, Fund. Math., 71 (1971), 273-283.

[5] ________, On the normal structure of automorphism groups of abelian p-groups, J. London Math. Soc., (2) 5 (1972), to appear.

[6] ________, The automorphism group of an abelian p-group and its non-central normal subgroups, to appear.

[7] A. Rosenberg, The structure of the infinite general linear group, Ann. of Math., 68 (1958), 278-294.

DEPARTMENT OF MATHEMATICS

UNIVERSITY OF HOUSTON

HOUSTON, TEXAS 77004

Countable Type Local Theorems In Algebra

Kenneth K. Hickin*

The well-known technique of Malcev provides a general set-theoretical scheme for proving theorems of the type: suppose enough finitely generated subalgebras of an algebra A have a property Σ; then A has Σ. In such proofs, one demands that A possess a "local system" of Σ subalgebras.

In this paper we attempt to provide set-theoretical techniques of similar generality for proving theorems of the similar type: suppose enough countable subalgebras of an algebra A have Σ; then A has Σ. Part of our problem is to formulate a suitable notion of "countable local system."

We consider three types of "countable local systems." Let $\underline{L}$ be a collection of subsets of a set S such that $\underline{L} = S$. Then (1) $\underline{L}$ is a C_1 system of S if $\underline{L}$ is countably directed, that is, for each sequence $K_1, \ldots, K_n, \ldots$ of members of $\underline{L}$, there is some $K \in \underline{L}$ such that $K_n \leq K$ for all n; (2) $\underline{L}$ is a C system of S if $\underline{L}$ is directed by inclusion, every member of $\underline{L}$ is countable, and whenever $K_1 \leq \ldots \leq K_n \leq \ldots$ are in $\underline{L}$, then $\overset{\infty}{\underset{n=1}{}} K_n \in \underline{L}$; (3) $\underline{L}$ is a W system of S if each member of $\underline{L}$ is countable and $\underline{L}$ has a member in common with each C system of S.

That every C system of S is a W system follows from the fact that the C systems of S form a filter base. Also, every W system is a C_1 system.

*This is a partial summary of a forthcoming paper of the same title.

Let $\Gamma = C_1$, C, or W. $\Gamma(\Sigma)$ denotes the class of all algebras possessing a Γ system of Σ subalgebras. Σ is called Γ-local if $\Gamma(\Sigma) \leq \Sigma$ and Σ is called Γ-downward if $\Sigma \leq \Gamma(\Sigma)$. Σ is Γ-closed if it is both Γ-local and Γ-downward.

If $s\Sigma = \Sigma$, then $C_1(\Sigma) = C(\Sigma) = W(\Sigma)$.

We shall use the concepts of C and W systems to develop a more general framework for proving countable local theorems.

Main Lemma. Suppose $\underline{W}$ is a W system of a set S and for each $K \in \underline{W}$, $f(K)$ is a finite subset of K. Then there is a W system $\underline{U} \leq \underline{W}$ of S such that f is constant on $\underline{U}$.

This generalizes a "regressive function theorem" of G. Fodor.

Some Applications

A. Direct Decompositions of Groups

Let Σ satisfy the conditions (x) (1) Σ is W-local, (2) every Σ group equals the normal closure of a finite subset, and (3) every Σ group is centerless. Let $DS\Sigma$ = class of direct sums of Σ groups, and $\underline{I}$ = class of indecomposable groups.

Theorem. Suppose Σ satisfies (x). Then $DS(\Sigma \cap \underline{I})$ is W-local.

This result shows that a centerless group must decompose into nice pieces if enough countable subgroups do.

Corollary. Let $\underline{R}$ = class of completely reducible groups without center, Σ_1 = centerless groups with Max-N, and Σ_2 = indecomposable, centerless, finitely generated groups. Then $\underline{R}$, $DS\Sigma_1$, and $DS\Sigma_2$ are W-local.

B. Cardinality of Automorphism Groups

If A is any algebra, the automorphic character of A is defined to be $\alpha(A)$ = the smallest cardinal number σ such that A has a subset S, $|S| = \sigma$, which is pointwise fixed only by the identity automorphism of A.

Lemma. The class of algebras having finite automorphic character is W-local

Theorem. The class of algebras A satisfying $|\text{Aut } A| \leq |A|$ is W-local, but not W-closed.

Final Remarks. D. W. Kueker has given a metamathematical characterization of C-closed classes. Many of our results can also be obtained by his methods.

I do not know if the concept of a W system is the optimum one for countable local theorems. In particular, can a regressive function lemma be obtained for a weaker type of system?

ON POWERS, CONJUGACY CLASSES AND SMALL-CANCELLATION GROUPS

SEYMOUR LIPSCHUTZ

TEMPLE UNIVERSITY

1. INTRODUCTION

Let P denote the following property of a group G:

[P] IF W in G has infinite order, then $W, W^2, W^3, \ldots$ are in different conjugacy classes.

Many important classes of groups have property P. For example,

(i) free groups,

(ii) fundamental groups of orientable 2-manifolds,

(iii) Fuchsian groups,

(iv) braid groups.

The author proved [5] that eighth-groups have property P. A similar argument extends this result to more general small-cancellation groups.

Theorem 1. Finitely presented sixth-groups and T-fourth-groups have property P.

Property P is also preserved under free products with cyclic amalgamations. That is:

Theorem 2. Let K_i be a class of groups with property P. Let G be the free product of the K_i with a cyclic group H amalgamated. Then G also has property P.

2. BASIC DEFINITIONS AND NOTATION

Let G be a finitely presented group with defining relations $R_1 = 1, \ldots, R_t = 1$. We assume without loss of generality that the relators R_i form a symmetric set, i.e. that the R_i are cyclically reduced and are closed under the operations of taking inverses and cyclic transforms. We consider the following two cancellation conditions on the relators R_i:

(i) Metric condition M(k). If R_i and R_j are not inverses, then less than 1/k of the length of either relator is cancelled in the product R_iR_j.

(ii) Triangle condition T. For any relators R_i, R_j and R_k, at least one of the products R_iR_j, R_jR_k, R_kR_i is freely reduced without cancellation.

We call G a sixth-group if G satisfies M(6), and we call G a T-fourth-group if G satisfies condition T and M(4). These groups are of interest because Greendlinger solved the word and conjugacy problems for them (c.f. [1] - [3]). Also, Lyndon [6] solved the word problem and Schupp [8] the conjugacy problem for classes of groups with similar conditions.

Capital letters V, W, A, B, ... will denote freely reduced words unless otherwise stated or implied. We use the notation:

$|W|$ is the word-length of W,

$V = W$ means V and W are the same element of the group G,

$V \cong W$ means V and W are identical words,

$V \sim W$ means V is conjugate to W.

We say W is fully reduced or R-reduced if W is freely reduced and does not contain more than half of a relator. We say W is cyclically fully reduced or cyclically R-reduced if every cyclic transform of W is fully reduced.

3. PROOF OF THEOREM 1

Let G be a sixth-group or a T-fourth-group. We need the following three lemmas about G. Here r is the length of the largest defining relator of G.

Lemma 1. Let W and V be R-reduced elements of G and suppose $W = V$. Then $|W| \leq r\,|V|$.

Lemma 2. Suppose W in G has infinite order. Then W^2 is conjugate to V where V and all its powers are cyclically R-reduced.

Lemma 1 is a restatement of Lemma 1 of [4], and Lemma 2 is a direct consequence of Lemma 2 of [4].

Lemma 3. Let W and V be cyclically R-reduced elements of G and suppose W is conjugate to V. Then $|W| \leq r^3\,|V|$.

Proof of Lemma 3. By the discussion on page 745 in [4], either $|W| \leq r$ or $W = TVT^{-1}$ where $|T| \leq r$. In the latter case, we use Lemma 1 to obtain

$$|W| \leq r\,|TVT^{-1}| \leq r(|T| + |V| + |T^{-1}|) \leq r(2r + |V|).$$

In either case, $|W| \leq r^3|V|$, as claimed.

Proof of Theorem 1. Suppose W in G has infinite order. By Lemma 2, W^2 is conjugate V where V and all its powers are cyclically R-reduced. Suppose $|m| > |n|$ and $W^m \sim W^n$. Then $W^{2m} \sim W^{2n}$, and so $V^m \sim V^n$. Say $V^m = TV^nT^{-1}$. Then

$$V^{m^2} = T\,V^{mn}\,T^{-1} = T^2\,V^{n^2}\,T^{-2}.$$

Similarly, $V^{m^4} \sim V^{n^4}$ and so fourth. In other words, for every positive integer s,

$$V^{m^{2^s}} \sim V^{n^{2^s}}.$$

But V and all its powers are cyclically R-reduced; hence, by Lemma 3,

$$|m|^{2^s}\,|V| = |V^{m^{2^s}}| \leq r^3\,|V^{n^{2^s}}| = r^3 |n|^{2^s}\,|V|$$

Therefore, for every positive integer s,

(1) $$|m|^{2^s} \leq r^3 |n|^{2^s}$$

But $|m| > |n|$ and r is a constant; hence (1) cannot hold for some large s. This is a contradiction, and so the theorem is proved.

4. PROOF OF THEOREM 2

Suppose W in G has infinite order, and suppose $W^m \sim W^n$. Then $V^m \sim V^n$ for a cyclically reduced conjugate V of W. There are two cases. (Here we write $\lambda(V)$ for the syllable (free product) length of V.)

Case 1. $\lambda(V) > 1$. Since V is cyclically reduced and $\lambda(V) > 1$, we have

$$\lambda(V^m) = |m|\lambda(V) \qquad \text{and} \qquad \lambda(V^n) = |n|\lambda(V).$$

Also, V^m and V^n are cyclically reduced. Since $V^m \sim V^n$, Solitar's theorem on conjugacy in free products with amalgamations (c.f. [7]) tells us that $\lambda(V^m) = \lambda(V^n)$. Therefore,

$$|m|\lambda(V) = \lambda(V^m) = \lambda(V^n) = |n|\lambda(V).$$

Accordingly, $|m| = |n|$.

Case 2. $\lambda(V) \leq 1$. Then V and V^m and V^n belong to a factor, say K, of G. Solitar's theorem tells us that there are two possibilities:

(1) $V^m \sim V^n$ in K. Since K has property P, we must have that $|m| = |n|$.

(2) There exist elements $h_1, h_2, \ldots, h_t$ in the amalgamated cyclic group H such that

$$V^m \sim h_1, \; h_1 \sim h_2, \; \ldots, \; h_{t-1} \sim h_t, \; h_t \sim V^n$$

where each conjugation takes place in a factor of G. Let h be a generator of the cyclic group H, and suppose

$$h_1 = h^{n_1}, \; h_2 = h^{n_2}, \; \ldots, \; h_t = h^{n_t}$$

Since the factors of G have property P,

$$|n_1| = |n_2| = \dots = |n_t|$$

Therefore,

$$v^m \sim h^{n_1} \sim v^n \quad \text{or} \quad v^m \sim h^{n_1} \sim v^{-n}$$

in K. Since K has property P, we obtain $|m| = |n|$. Thus the theorem is proved.

5. PROBLEMS

We list a few problems.

(1) Let G have property P and let E be a finite extension of G. Does E have property P?

(2) Let G be the free product of two free groups with a finitely generated group amalgamated. Does G have property P?

(3) Give other classes of groups with property P.

(4) Give an algebraic proof of the fact that the braid groups have property P. (All known proofs are topological.)

(5) Show that other small-cancellation groups (c.f. Lyndon [6] and Schupp [8]) have property P.

BIBLIOGRAPHY

[1] Greendlinger, M., Sibirsk. Mat. Z. 7, 785-803 (1966)

[2] ________, Doklady Akad. Nauk SSSR 154, 507-509 (1964)

[3] ________, Comm. Pure Appl. Math. 13, 641-677 (1960)

[4] Lipschutz, S., Comm. Pure Appl. Math. 23, 743-747 (1970)

[5] ________, Arch. Math. 23, 121-124 (1972)

[6] Lyndon, R. C., Math. Ann. 166, 208-228 (1962)

[7] Magnus, W., Karass, A., and Solitar, D., Combinatorial group theory, Wiley, New York, 1965

[8] Schupp, P. E., Math. Ann. 178, 119-130 (1968)

The Orders of Generation of SL(2,R)

F. Lowenthal*

A connected Lie group H is generated by a pair of one-parameter subgroups if every element of H can be written as a finite product of elements chosen alternately from the two one-parameter subgroups, i.e., if and only if the subalgebra generated by the corresponding pair of infinitesimal transformations is equal to the whole Lie algebra h of H. If, moreover, there exists a positive integer n such that every element of H possesses such a representation of length at most n, then H is said to be uniformly finitely generated by the pair of one-parameter subgroups. In this case, define the order of generation of H as the least such n; otherwise define it as . Since the order of generation of H will, in general, depend upon the pair of one-parameter subgroups, H may have many different orders of generation. However, it is a simple consequence of Sard's theorem that the order of generation of H must always be $\geq$ the dimension of H.

The one-parameter subgroups of SL(2,R) can be expressed in the form:

(a) $A(t) = C\begin{pmatrix} \cos t & \sin t \\ -\sin t & \cos t \end{pmatrix}C^{-1}$, $C \in SL(2,R)$

(b) $A(t) = C\begin{pmatrix} 1 & t \\ 0 & 1 \end{pmatrix}C^{-1}$, $C \in SL(2,R)$

(c) $A(t) = C\begin{pmatrix} e^{t} & 0 \\ 0 & e^{-t} \end{pmatrix}C^{-1}$, $C \in SL(2,R)$

depending upon whether it is elliptic, parabolic, or hyperbolic, respectively. Note that a one-parameter subgroup is compact if and only if it is elliptic.

---

*A full discussion of these results will appear soon in the Canadian Journal of Mathematics.

The only two dimensional connected Lie subgroups of SL(2,R) are those that leave a one-dimensional subspace of R^2 invariant. Hence a pair of distinct one-parameter subgroups of SL(2,R) generate SL(2,R) if and only if their infinitesimal transformations do not have a common eigenvector. A pair of distinct infinitesimal transformations of SL(2,R), i.e., elements of the Lie algebra sl(2,R), with no common eigenvector can be simultaneously transformed into one and only one of 7 possible normal forms by means of a suitably chosen inner automorphism of SL(2,R); these 7 forms correspond to the following cases: (a) both elliptic; (b) one elliptic, one parabolic; (c) one elliptic, one hyperbolic; (d) one parabolic, one hyperbolic; (e) both parabolic; (f) both hyperbolic with eigenvectors interlacing; (g) both hyperbolic with eigenvectors separating. The order of generation of SL(2,R) can be determined in each case; the proof and computations are partly general and partly peculiar to the particular case. The result is the following theorem.

Theorem: If SL(2,R) is generated by a pair of one-parameter subgroups, then the order of generation is 3, 4, 8 or ∞. It is ∞ if both are elliptic, 3 if exactly one is elliptic, and 4 in all other cases except that it is 8 if both are hyperbolic with interlacing eigenvectors.

Centrally Separated Elements in Groups

Richard E. Phillips

I. Recall the definitions of the classes of groups Z, SI and SN ([5, p. 182 and 218] or [6, p. 139]). Thus, G is a Z-group if G possesses a central system, an SI-group if G has an invariant system with abelian factors, and SN-group if G has a normal system with abelian factors. We note some elementary properties of these three classes:

1.1 $Z \subset SI \subset SN$

1.2 all three are subgroup closed,

1.3 all three are closed under the taking of cartesian products,

1.4 all three satisfy the local theorem ([5, pp. 183-189] or [6, pp. 136-140]).

II. The concept of a system of subgroups of a group is a rather awkward one. It is our intent here to give characterizations of the above three classes which will not use tne notion of a system of subgroups. To this end, we need some definitions.

Definition 1. A finite non-empty subset F of $G^{\#}(G^{\#} = G - \{1\})$ is centrally separated in G if there is a positive integer n and subgroups

$$A_n \subsetneqq B_n \subset A_{n-1} \subsetneqq B_{n-1} \subset \dots \subset A_1 \subsetneqq B_1$$ of G such that:

i) for each i, B_i/A_i is a central factor of G, and

ii) $F \subset \bigcup_{i=1}^{n} (B_i \setminus A_i)$

Definition 2. Let Σ be a variety of groups. A finite non-empty subset F of $G^{\#}$ is Σ-invariantly (Σ-normally) separated if there is a positive integer n and subgroups

$$A_n \underset{\neq}{\subset} B_n \subset A_{n-1} \underset{\neq}{\subset} B_{n-1} \subset \dots \subset A_1 \underset{\neq}{\subset} B_1 \text{ of } G \text{ such that:}$$

i) For each i, A_i and B_i are normal in G ($A_i \lhd B_i$)

ii) for each i, $B_i/A_i \in \Sigma$

iii) $F \subset \bigcup_{i=1}^{n} (B_i \setminus A_i)$

Our first theorem gives a somewhat different characterization of the separation properties.

Theorem 1 (K. Hickin). Let F be a finite non-empty subset of $G^{\#}$.

i) F is centrally separated if and only if for every $\phi \neq M \subset F$, $M \not\subset [\langle M\rangle, G] = [M,G]^G$.

ii) F is Σ-invariantly separated if and only if for every $\phi \neq M \subset F$, $M \not\subset \Sigma(M^G)$. (here $\Sigma(M^G)$ is the verbal subgroup of M^G).

iii) F is Σ-normally separated if and only if for every finite $\phi \neq M \subset F$, $M \not\subset \Sigma(M)$.

Theorem 2.

i) G is a Z-group if and only if for every finite non-empty subset $F \subset G^{\#}$, F is centrally separated.

ii) G is a $\Sigma I(\Sigma N)$-group if and only if for every finite non-empty subset $F \subset G^{\#}$, F is Σ-invariantly (Σ-normally) separated.

Combining the previous two theorems, we have:

Corollary

i) G is a Z-group if and only if for every finite non-empty subset $F \subset G^{\#}$. $F \not\subset [\langle F\rangle, G]$.

ii) G is an SI-group if and only if for every finite non-empty subset $F \subset G^{\#}$, $F \not\subset (F^{G})'$

iii) G is an SN-group if and only if for every finite non-empty subset $F \subset G^{\#}$, $F \not\subset \langle F\rangle'$.

Recall that a subgroup H of a group G is serial in G if H is a member of some normal system of G. Using the techniques employed above, we are able to show:

Theorem 3. H is a serial subgroup of G if and only if for every finite subset F of G such that $F \not\subset H$, $F \not\subset H^{F}$.

III. The characterization of Z groups given in the Corollary (i) admits the following generalizations.

Definition 3.1 Let n be a positive integer. Define Z(n) to be the class of all groups G such that for every non-empty subset $F \subset G^{\#}$ with $|F| \leq n$, $F \not\subset [\langle F\rangle, G]$.

SI-groups admit similar generalizations, but these will not be developed here.

For the sake of a consistent terminology, define $Z(\infty) = Z$. The following facts can easily be verified.

3.1 $Z = Z(\infty) \subset \ldots \subset Z(n) \subset \ldots \subset Z(2) \subset Z(1)$

3.2 for each n, $1 \leq n \leq \infty$, $Z(n)$ is closed under the taking of subgroups and cartesian products.

3.3 for each n, $1 \leq n \leq \infty$, $Z(n)$ satisfies the local theorem.

$Z(1)$ groups were introduced by Durbin in [2] and [3], where they are called residually central. The papers [1] and [7] also deal with $Z(1)$-groups. In [2], Durbin asks whether $Z(1) = Z$. This question appears to be difficult. The following are some related questions.

Problem 1. Is $Z(n) = Z$ for any n?

Problem 2. Is $Z(n) \subset SN$ for any $n < \infty$?

In [2], it is shown that for groups with min-N, $Z(1) = Z$. It follows from the local theorems that within the class of locally finite groups, $Z(1) = Z$. It would be interesting to investigate the relationship between $Z(1)$ and Z in groups subjected to various other conditions. In particular, we pose:

Problem 3. Is every solvable $Z(1)$ group a Z group?

In this direction, we have two theorems.

Theorem 4. Every supersolvable $Z(1)$ group is a Z group.

Theorem 5. (J. E. Roseblade) Every "abelian by nilpotent" $Z(1)$ group is a Z group.

Finally, for the purpose of constructing examples, it would seem worthwhile to determine which wreath products are in the classes $Z(n)$, $1 \leq n \leq \infty$.

Problem 4. Find necessary and sufficient conditions on A and B so that A wr B is in $Z(n)$.

The work of Hartley in [4] is probably related to Problem 4.

Bibliography

1. C. Ayoub, On properties possessed by solvable and nilpotent groups, J. Austr. Math. Soc. 9(1969), 218-227.
2. J. Durbin, Residually central elements in groups, J. of Algebra 9(1968), 408-413.
3. J. Durbin, On normal factor coverings of groups, J. of Algebra 12(1969), 191-194.
4. B. Hartley, The residual nilpotence of wreath products, Proc. London Math. Soc. (3) 20(1970), 365-392.
5. A. G. Kurosh, Theory of Groups, Vol. II, (1956). Chelsea, New York.
6. D. J. S. Robinson, Infinite solvable and nilpotent groups, Queen Mary College Math. notes (1968).
7. O. Slotterbeck, Finite factor coverings of groups, J. of Algebra 17(1971), 67-73.

AN ATTACK ON THE RESTRICTED BURNSIDE PROBLEM FOR GROUPS OF EXPONENT 8 ON 2 GENRATORS

Ricardo B. Quintana, Jr.

We denote by $B(n,r)$ the freest group of exponent n on r generators and we let $b(n,r)$ be an integer such that every finite group $B(n,r)$ has order less than or equal to $b(n,r)$. The Restricted Burnside Problem is, given a finite group $B(n,r)$, to prove $b(n,r)$ exists and if possible to produce it. Here I want to outline a method for producing $b(8,2)$. However, before I do so, I shall have to explain what Kostrikin did in [2] in showing $b(5,2) \leq 5^{34}$.

Consider $B(5,2)$ and let L' be its associated Lie ring. Then L' is a Lie algebra over $GF(5)$. Also L' satisfies the 4th Engel condition by a result of Higman's. Let $K=L/I$ where I is the ideal generated by the elements $[uv^4]$, $u, v \varepsilon L$ and L is the free Lie ring generated by x and y. Then L' is a homomorphic image of K. Kostrikin produced a spanning set for the quotient rings K_i/K_{i+1} $(i=2,3,\ldots,13)$ of the descending central series of K and showed $K_{13}=K_{14}$. To do so, he started with a known spanning set, i.e., basic Lie elements of degree i. The basic Lie elements correspond in the Lie ring to simple commutators of length i in the group which, of course, form a spanning set for $B(5,2)_i/B(5,2)_{i+1}$. Kostrikin then determined all the relations that existed among simple elements of the same degree. To determine these relations, he started with

(1) $[P(y+zQ)^4] \equiv o(\text{mod } I)$

where P and Q are Lie monomials, y is a generator, and $z \in GF(5)$. He then obtained

(2) $[P(y+zQ^4] \equiv b(P,Q^\circ)+zb(P,Q')+z^2b(P,Q^2)+z^3b(P,Q^3)$
$+z^4b(P,Q^4)$

and showed by choice of z that

(3) $b(P,Q^\circ) \equiv b(P,Q') \equiv b(P,Q^2) \equiv b(P,Q^3) \equiv b(P,Q^4) \equiv o \text{ mod } I.$

The relations

(4) $b(P,Q') = [Py^3Q]+[Py^2Qy]+ [PyQy^2] +[PQy^3] \equiv o(I)$

(5) $b(P,Q^2) = [Py^2Q^2] + [PyQyQ] + [PyQ^2y] + [PQ^2y^2]$
$+[PQyQy] + [PQy^2Q] \equiv o\ (I)$

are those that are relevant in producing a basis for K_i/K_{i+1}. Simple Lie monomials are then written as f^k_{ij} where i = the total degree in x and y, j = the degree in s, and k = an index used for the purpose of numbering monomials. By convention, he began with x. For example, $f^1_{5,1} = [xy^4]$. Here as in (4) and (5), the bracket is used to denote the Lie product as opposed to ordinary multiplication. Let p_i be the number of linearly independent elements in K_i/K_{i+1}. By $\overline{f^k_{ij}}$ we mean f^k_{ij} with x and y interchanged, e.g., $\overline{f^1_{5,1}} = [yx^4]$. Because of symmetry of f^k_{ij} in x and y, we examine only those monomials for which $j \leq [\frac{i}{2}]$. From the definition of I, we have

(6) $[xy^4] \equiv o \text{ mod } I$

Also, it follows from the definition of multiplication in the Lie ring that

$$(7)\quad [xy^2x] = -[yx^2y]$$
$$[xy^2x] = [xyxy]$$
$$[yx^2y] = [yxyx]$$

Starting with K_5/K_6, we have

$$f^1_{5,1} = [xy^4] \equiv o(I)$$
$$f^1_{5,2} = [xy^2xy]$$
$$f^2_{5,2} = [xy^3x] = 2f^1_{5,2}(I)$$
$$f^3_{5,2} = [xyxy^2] = [xy^2xy] = f^1_{5,2} \text{ by } (7)$$

The expression for $f^2_{5,2}$ is obtained from (4) as follows:

$$b(x,x) = [xyxy^2] + [xy^2xy] + [xy^3x]$$
$$= f^2_{5,2} + \underline{2f^1_{5,2}} \equiv o \bmod I$$

Hence $p_5 \leq 2(f^1_{5,2}, f^1_{5,2})$.

Continuing in this fashion, Kostrikin showed $p_6 \leq 4$, $p_7 \leq 4$, $p_8 \leq 4$, $p_9 \leq 6$, $p_{10} \leq 3$, $p_{11} \leq 2$, $p_{12} \leq 1$ and $K_{13}=K_{14}$

Now in the case of $G = B(8,2)$, we do not know that the Lie ring of the group satisfies the 7th Engel conditon. We therefore have to resort to methods other than Lie ring methods in order to produce a bound on the order of the freest finite group of this type. I shall now outline an approach that I am trying. Let $N(a)$ be the smallest normal subgroup of B(8,2) containing a generator a. For $x \varepsilon$ B(8,2),

$$(8)\quad 1 = x^{-8}(xa)^8 = x^{-8}xaxaxaxaxaxaxaxa$$
$$= a^{x^7}a^{x^6}a^{x^5}a^{x^4}a^{x^3}a^{x^2}a^{x}a^{1}$$
$$a^{x^7+x^6+x^5+x^4+x^3+x^2+x+1} \bmod N(a)'$$

where the addition is just the sum of endomorphisms in the ring of endomorphisms of a multiplicative abelian group. We imbed G homomorphically into the group of automorphisms of N(a)/N(a)' for each such a as follows:

Map $x \varepsilon G$ into $\bar{x}$ εAuto (N(a)/N(a)') where $\bar{x}$ is defined by

$$(kN(a)')\,\bar{x} = k^x N(a)',\ K\varepsilon N(a).$$

The mapping $x \to \bar{x}$ induces a mappingΘ: $\Sigma n_i x_i \to \Sigma n_i \bar{x}_i$ $n_i \varepsilon Z$, $x_i \varepsilon G$ of ZG (= the group ring of G = B(8,2) over the integers Z) into the ring of endomorphisms of N(a)/N(a)'. We note that if $s_8(G)$ is defined to be the ideal of ZG generated by the elements

$$1 + x + x^2 + x^3 + x^4 + x^5 + x^6 + x^7,\ x\varepsilon G,$$

then $S_8(G) \subseteq$ kernel (Θ). We now work in ZG mod $S_8(G)$. The defining congruence for $S_8(G)$ is

$$(9)\ 1 + x + x^2 + x^3 + x^4 + x^5 + x^6 + x^7 \equiv o \bmod S_8(G)$$

for all $x\ \varepsilon G$. Let $x = 1 + X$. Then in ZG,

$$(10)\ 4X^4 \equiv 2X^8 \equiv o \bmod S_8(G)$$

and

$$(11)\quad 4X + 2X^3 + X^7 \equiv o \bmod S_8(G).$$

Note: by collecting $1 = a^{x^7}a^{x^6}a^{x^5}a^{x^4}a^{x^3}a^{x^2}a^{x}a^{1}$ via the Hall collection process, we have

(12) $(a,x)^4(a,x,x,x)^2(a,x;7) = C_3^4 C_4^2 C_8^1$

where C_j^i = a product of commutators of length i or more each raised to the jth power. Now we want in ZG a relation that corresponds to $[uv^4] \equiv \text{mod } I$ in L/I of exponent 5,2 generator work. To this end we multiply (11) by X^5 and use (10) to produce

(13) $X^{12} \equiv 0 \bmod S_8(G)$.

We must, in order to show $ZG_{n_0} = ZG_{n_{0+1}}$ for some positive integer n_0, determine as many as possible, and hopefully all, of the relations that exist among simple group ring elements of the same degree. To this end, we start with

(14) $[(1+X)(1+A)^n-1]^{12} \equiv b(X^{12},A^0) + nb(X^{11},A^1) + n^2b(X^{10},A^2) + \ldots + n^{11}b(X^1,A^{11}) + n^{12}b(X^0,A^{12}) \bmod J_{13}$

where J_i is the ideal generated by $S_8(G)$ and all products of monomials in A and X of degree i, and

(15) $b(X^{12},A^0) = X^{12}$

$b(X^{11},A^1) = X^{11}A + X^{10}AX + \ldots + AX^{11}$

etc.

By choice of n,

(16) $b(X^{12},A^{0}) \equiv b(X^{11},A^{1}) \equiv \ldots \equiv b(X^{0},A^{12}) \equiv o \bmod J_{13}$.

Now in ZG, we want to work with monomials that correspond in G to simple commutators. For example, simple commutators of length 5 are

(a,x,x,x,x), (a,x,x,a,x), (a,x,x,x,a), (a,x,a,x,x)

and those obtained by interchanging a and x. In ZG_4 mod J_5, we have

$$f^{1}_{4,1} = X^4$$
$$f^{1}_{4,2} = X^2AX$$
$$f^{2}_{4,2} = X^3A$$
$$f^{3}_{4,2} = XAX^2$$

corresponding to these simple commutators. Simple group ring elements are here denoted by f^{k}_{ij} where i denotes the total degree in X and A, j-1 denotes the degree in A and k is an index used for the purpose of numbering monomials. Note that the degree in A is j-1 since $f^{1}_{4,1}$ for example corresponds to the simple commutator (a,x,x,x,x). Thus if S_i is the number of linearly independent elements in ZG_i/J_{i+1}, then $S_4 \leq 8$ $(f^{1}_{4,1}, \overline{f^{1}_{4,1}}, f^{1}_{4,2}, \overline{f^{1}_{4,2}}, f^{2}_{4,2}, \overline{f^{2}_{4,2}}, f^{3}_{4,2}, \overline{f^{3}_{4,2}})$ where as before, $\overline{f^{i}_{jk}}$ is the result of interchanging X and A in f^{i}_{jk}. We continue simply counting group ring elements corresponding to simple commutators until we come to $ZG_{14} \bmod J_{15}$. Now relations (16) could be applied to $ZG_{12} \bmod J_{13}$. However, we wish to reinterpret relations in ZG as relations in B(8,2). By multiplying the relations in (15) by a monomial on the left of degree 2 we produce relations which can be so interpreted. For example, $X^{12} \equiv o(S_8)$ came from (11) by multiplication by X^5 on the left, i.e., $4X^6 + 2X^8 + X^{12} \equiv o\ (S_8)$. But $(a,x;4)^4 = C_5^1$ so $(a,x;6)^4 = C_{11}^1$. However, in (2) it is proved that

(17) $(a_3,x;8)^2 \equiv 1 \bmod G_{16}$ for $a_3 \in G_3$.

Hence a relation such as $XAX^{12} \equiv o\ (J_{15})$ can be reinterpreted as $(a,x,a,x;12) \equiv 1 \bmod G_{16}$. We write $XAX^{12} \equiv o\ (J_{15}^*)$ where * means that such a reinterpretation in G is possible. Hence we begin to use relations (16) in analyzing ZG_{14} mod J_{15}^*, i.e.,

$$f_{14,1}^1 = X^{14} \equiv o\ (J_{15}^*)$$
$$f_{14,2}^1 = X^{13}A$$

etc.

Our hope is that after multiplying relations (16) by a monomial of degree 2 on the left we can, by substituting x^2 for x, x^4 for x, a^2 for a, a^4 for a, and further multiplication on either the left or right, produce sufficiently many relations so as to be able to show $ZG_{n_0} \equiv o\ (J_{n_0+1}^*)$ for some positive integer n_0. By so doing, we would answer the restricted Burnside conjecture for B(8,2) in the affirmative and produce a b(8,2).

Bibliography

1. Kostrikin, A. I., "Solution of the Restricted Burnside Problem for the Exponent 5," Izvestia Akad. Nauk. SSSR, Ser. Mat. 19 (1955), 233-244.

2. Krause, E. F., Ph.D. thesis, University of Wisconsin.

Groups whose images have a transitive normality relation

Derek J. S. Robinson

1. Introduction

If P is a group theoretical property, a *just non-P group* (or *JNP-group*) is a group which does not have P but whose proper images do. JNP-groups have been studied by Newman ([3], [4]) when P is commutativity and by Mann ([2]) and Weichsel ([6]) when P is regularity. If one knows sufficient about groups with P, it may be possible to determine all the soluble JNP-groups: here Newman's work is a case in point. The hypothesis of solubility is made partly to guarantee the existence of normal subgroups: however one may still be able to say something in the insoluble case by utilizing Fitting's theory of semisimple groups.

Let T denote the property "normality is transitive": thus G is a *T-group* if $H \lhd K \lhd G$ always implies that $H \lhd G$. Soluble T-groups have been studied by several authors and can be regarded as fairly well-known: see for example [1] and [5]. If one regards T as a generalization of commutativity it is natural to ask if all soluble JNT-groups can be classified. It turns out that this can indeed be done: the classification, as one might expect, is rather more complicated than Newman's: for example one finds nine distinct types of groups as opposed to his two.

2. An application

Before describing the classification we shall indicate an application. If one examines the list of soluble JNT-groups, one soon observes that no group can be both finitely generated and infinite. Thus a finitely generated soluble JNT-group is finite. Now let G be a finitely generated soluble group which is not a T-group, and suppose that $\{N_\alpha : \alpha \in A\}$ is a chain of normal subgroups of G such that no G/N_α is a T-group. If U is the union of the chain, then G/U is not a T-group: this stems from the fact that a finitely generated soluble T-group is finite or abelian ([5]), and hence is certainly finitely presented. By Zorn's Lemma there is an $N \lhd G$ maximal subject to G/N not being a T-group. Then G/N is a JNT-group, whence it is finite. We have therefore proved the following:

A finitely generated soluble group which is not a T-group possesses a finite image which is not a T-group.

3. The classification

Let G be a soluble JNT-group. Should G happen to be nilpotent, every proper image will be abelian or hamiltonian, by the Dedekind-Baer Theorem on groups with every subgroup normal. In fact one can exclude the possibility of a hamiltonian image by a simple argument. This being so, we are left with the nilpotent just non-abelian groups (other than the quaternion group of order 8) and these are all on Newman's list. Briefly the structure of

G is this: G is a p-group, G' has order p and lies in the center Z; moreover Z is either cyclic or quasicyclic and G/Z is elementary abelian. G/Z can therefore be regarded as a vector space over GF(p) and commutation is readily seen to give rise to a non-degenerate skew-symmetric bilinear form on G/Z. In a word these groups (Type I in our classification) have a rather simple structure.

Turning to the non-nilpotent case, we write

$$L = [G',G] ,$$

a subgroup which is the last term of the lower central series of G since the class of a nilpotent T-group is at most 2.

Obviously an important role will be played by the minimal normal subgroups of G since their factor groups contain maximal information. But it is conceivable that G has no such subgroups and it is this case which we shall examine in greater detail.

With the assumption that G has no minimal normal subgroups one proves successively that

(i) L is torsion-free and abelian,

(ii) L is rationally irreducible as a G-module,

(iii) L is divisible.

In proving these, crucial use is made of the existence of an infinite descending chain of normal subgroups of G within L having trivial intersection. Step (iii) is the most troublesome.

Next one shows that G splits over L. Let X be a complement of L: it turns out that X is abelian. By (i)-(iii) L is an irreducible QX-module, Q being the rational field. Choose $a \neq 1$ from L: the QX-homomorphism $r \longrightarrow a^r$ yields an exact sequence

$$0 \longrightarrow K \longrightarrow QX \longrightarrow L \longrightarrow 0 ;$$

here K is a maximal ideal of the commutative ring QX.

Since L is not minimal normal in G and yet is rationally irreducible, there exists a prime p and a normal subgroup P of G such that $1 < P < L$ and L/P is a p-group. Naturally L/P is a divisible abelian p-group and each of its subgroups is normal in G/P. Consequently an x in X induces in L/P a power automorphism, mapping each element to a power. This automorphism may be described by a p-adic integer unit α_x where

$$(aP)^x = (aP)^{\alpha_x} = (aP)^n,$$

α_x being congruent to n modulo the order of aP. Moreover α_x does not depend on the subgroup P but merely on the prime p.

Thus we have a homomorphism $x \longrightarrow \alpha_x$, which can be extended in the obvious way to a ring homomorphism $\alpha: QX \longrightarrow F_p$ where F_p is the field of p-adic numbers. One verifies that $\text{Ker}\,\alpha = K$ so that, on setting $F = \text{Im}\,\alpha$, one obtains a second exact sequence

$$0 \longrightarrow K \longrightarrow QX \longrightarrow F \longrightarrow 0 .$$

Thus the X-module L is isomorphic with the $Y = X^{\alpha}$-module F and F is a subfield of F_p.

Next let $A = a^X$: then G/A is a T-group, so that for any integer n and x in X

$$(a^{\frac{1}{n}} A)^x = (a^{\frac{1}{n}} A)^m$$

for a suitable integer m. This yields the equation

$$QX = Q + Rg\langle X\rangle + K.$$

On applying α we get

$$F = Q + Rg\langle Y\rangle .$$

It is now possible to construct G explicitly. Let p be a prime, F a subfield of F_p and X a group of p-adic integer units in F, other than $\langle -1\rangle$, such that

$$Rg\langle X\rangle < Q + Rg\langle X\rangle = F .$$

G is the natural split extension of F by X. This is Type IX in our classification.

The remaining seven types all have minimal normal subgroups: we shall briefly indicate their nature.

Types II and III are infinite 2-groups. Essentially the most complicated group which can arise is constructed as follows. Let L be a group of type 2^∞ with canonical generators $a_1, a_2, \ldots,$ let $D = \langle d\rangle$ be a group of order 1 or 2 and let X be an extra-special 2-group with generators of order 2. Denote by C the direct product of $L\times D$ with X in which $\langle a_1\rangle$ is amalgamated with the centre of X. Choose $\sigma \in Hom(X,\langle a_1\rangle)$. Define

$$G = \langle C, t\rangle$$

where

$$a^t = a^{-1}, \ (a\in L), \ d^t = a_1 d \ (\text{if } d \neq 1)$$

and

$$x^t = x^{1+\sigma}, \quad (x \in X);$$

in addition require that

$$t^2 = 1 \quad \text{or} \quad a_2 d \ (\text{if } d \neq 1).$$

A group of Type IV is merely the split extension of an elementary abelian group of order p^2 (with p an odd prime) by a diagonal but non-scalar subgroup of $GL(2,p)$. This is the only type of soluble JNT-group with more than one minimal normal subgroup (there are two).

Type V is the split extension of an extra-special p-group P of exponent $p > 2$ by an automorphism group $\langle t \rangle$ defined by $x_\lambda{}^t = x_\lambda{}^n$ for some $1 < n < p$: here $\{x_\lambda : \lambda \in \Lambda\}$ is a basis for P modulo its centre.

Types VI and VII are non-periodic groups requiring non-split extensions in their construction: they are too complicated to describe here. However the remaining Type VIII is perhaps the most obvious. One merely takes a soluble T-group X for which there is a faithful irreducible X-module M (not cyclic as an abelian group). Define G to be the natural split extension of M by X. Here M is fairly obviously the unique minimal normal subgroup of G. Of course one wishes to know for which groups X such a module exists. For example it may be shown that a finite soluble T-group X has a faithful irreducible module if and only if the centre of the Fitting subgroup of X is cyclic.

References

[1] W.Gaschütz, Gruppen in denen das Normalteilersein transitiv ist. J. reine angew. Math. 198, 87-92(1957).

[2] A. Mann, Regular p-groups, Israel J. Math. 10, 471-477(1971).

[3] M. F. Newman , On a class of metabelian groups. Proc. London Math. Soc. (3)10, 354-364(1960).

[4] M. F. Newman, On a class of nilpotent groups. ibid., 365-375.

[5] D.J.S. Robinson, Groups in which normality is a transitive relation. Proc. Cambridge Philos. Soc. 60, 21-38(1964).

[6] P. M. Weichsel, Just-irregular p-groups, Israel J. Math. 10, 359-363(1971).

POLYCYCLIC GROUP RINGS AND THE NULLSTELLENSATZ

J.E. ROSEBLADE

Jesus College, Cambridge, England

1. In his talk, Karl Gruenberg explained how a knowledge of rings could help when investigating certain sorts of soluble group, and he took us through some of the ring theory and some of the applications. I don't want to talk about group theoretic applications at all but to concentrate upon the ring theory. It will help if I recall for you the famous theorems of Philip Hall in his three papers in the Proceedings of the London Mathematical Society in 1954, 1959 and 1961 in as far as they relate to the ring theory. First there is the theorem of the 1954 paper.

> 1954: If H is polycyclic by finite and J commutative and Noetherian then S = JH has Max-r.

This is the modern version of Hilbert's Basis Theorem, and I mention it only to emphasize that we are dealing throughout with right Noetherian rings. H will always be polycyclic by finite and J will always be commutative and Noetherian. Indeed throughout the talk any unexplained letter will mean what it meant on its first appearance.

The second result he proved in 1959. Group theoretically it said that finitely generated Abelian by nilpotent groups were residually finite; ring theoretically the important discussion concerned irreducible representations of finitely generated nilpotent groups.

> 1959: If N is finitely generated nilpotent by finite then irreducible $\mathbb{Z}N$-modules are finite.

He did not restrict himself to integer coefficients, however; he also discussed irreducible representations of N over fields. His result in this direction is

> 1959: If K is a field all of whose non-zero elements are roots of unity then the simple KN-modules are all finite-dimensional over K.

The fields here are precisely those which are of prime characteristic and are absolutely algebraic; since they will play some rôle in to-day's talk I need a short word for them. I shall call them <u>absolute</u> fields. The best way to view these results is as a modern version of the so-called Weak Nullstellensatz. You will recall that this states that any simple module for a finitely generated algebra over a field is finite-dimensional; and that a first corollary is that simple modules for integral polynomial rings are finite. The difference is that the Weak Nullstellensatz is universally true, whatever the field of coefficients, and Hall's result is not. The only polycyclic by finite groups which have every irreducible module over any field finite-dimensional are the Abelian by finite ones, and this result is a very easy consequence of the classical result.

In that second paper he did discuss representations of non-nilpotent polycyclic groups, and I shall say more of that later. Just now I need point out only that he did not manage to prove either of these stated theorems for polycyclic by finite groups. Whether the simple $\mathbb{Z}H$-modules have all to be finite has come to be known as Philip

Hall's Problem.

The third result was about centralizers of chief factors of finitely generated Abelian by nilpotent groups. Group theoretically it said that in such a group the only elements which centralized every chief factor were those in the Fitting radical. The relevant ring theoretic result was somewhat stronger, and I will state it as

> 1961: If M is a finitely generated $\mathbb{Z}N$-module, with killer M^*, then any element central in $\mathbb{Z}N/M^*$ which kills all the simple images of M acts nilpotently upon M.

Taking M to be just a ring image of $\mathbb{Z}N$, this becomes

> If $(\mathbb{Z}N)^\phi$ is a ring image of $\mathbb{Z}N$ then any central element of $(\mathbb{Z}N)^\phi$ lying in the Jacobson radical is nilpotent.

As Hall pointed out himself, this corresponds to the classical Strong Nullstellensatz; and his proof corresponds to that given by Rabinowicz. In the paper he doubted whether this result would go over to the group algebra case in general, because of the failure of the earlier result for non-absolute fields; but he does not seem to have discussed the question of whether it does if $\mathbb{Z}$ is replaced by an absolute field. As before, he left open the question of whether the results would go over for polycyclic by finite groups.

I have been thinking for some time about these questions, and I should like to discuss some of the theorems which can be proved. By far the most difficult of the questions is whether the simple

$\mathbb{Z}H$-modules are finite, and I am still very much in the dark about that.[1] The question, after some manoeuvring, throws up a host of allied questions in number theory and algebraic geometry and it seems to me now that it is hardly group theory or ring theory any more. The trouble is that one is dealing, on the whole, with maximal right ideals of S in a distinctly non-commutative situation. To get over this one needs to discuss situations where there is a sufficiency of symmetry between right and left. To begin with, therefore, I shall concentrate upon two-sided ideals of $\mathbb{Z}H$, and modules of the form $(\mathbb{Z}H)^{\phi}$ for a ring homomorphism ϕ . We want a strong and a weak Nullstellensatz for S = JH in an ideal version.

2. The weak version concerns capitals of a ring. A capital of a ring X being merely a simple ring which is a homomorphic image of X. In other words they are the quotients X/Y, where Y is a maximal ideal of X: the things at the top of X. The ideal version of Philip Hall's problem would ask whether the capitals of $\mathbb{Z}H$ are all finite. I don't want, however, to restrict attention to integral coefficients; because it is not necessary to do that, and because it would hardly allow us to deal with the algebra version of the 1961 result. The capitals of $\mathbb{Z}$ are the finite prime fields. The capitals of K are just K. What these have in common is that they are both absolute. So what I shall assume throughout is that the capitals of J are absolute. These rings J are just the right ones, so I shall call them absolutely capital.

Now the strong ideal version of the Nullstellensatz will deal with ring images S^{ϕ} of S and will state, if true, that the Jacobson radical of S is nilpotent. As Gruenberg mentioned in his talk, the coefficient ring J, is always an image of JH, so that to prove a Strong Nullstellensatz for S it is necessary to assume that the

1) see the note at the end.

Strong Nullstellensatz holds already for J. I shall assume throughout, therefore, that if J^ϕ is an image of J then $Jac(J^\phi)$ is nilpotent. This is surely true for $J = \mathbb{Z}$ and $J = K$. The rings which satisfy this are called Jacobson rings. Throughout most of the talk, therefore, we shall be dealing with absolutely capital Jacobson rings.

Before we carry on let me remind you of Krull's version of the Nullstellensatz, which he discussed in the Proceedings of the International Congress 1950:

> If J is an absolutely capital Jacobson ring then so also is J[X], and, consequently, so is $J_1 = J\langle x\rangle$, the group ring of an infinite cyclic group $\langle x\rangle$ over J.

Now let me state the theorem and then discuss it.

Theorem. Suppose J is an absolutely capital Jacobson ring and H any polycyclic by finite group. Any capital of S = JH is finite-dimensional over some capital of J. If S^ϕ is a ring image of S, then $Jac(S^\phi)$ is nilpotent.

If we put $J = \mathbb{Z}$ here in the weak form, we see that capitals of $\mathbb{Z}H$ are all finite; so that Philip Hall's problem is true in the ideal version. If you like, this shows that to solve Hall's problem, all that has to be done, is to show that any maximal right ideal of S contains a maximal ideal of S. If we put J = K in the strong form then we get the algebra version of the 1961 result for absolute fields, but it remains problematic whether it is true for non-absolute fields.

3. Let me write $G = H \times \langle x\rangle$ for the direct product of H and $\langle x\rangle$;

and write $R = JG$ for its group ring over J. The idea which makes the proof work is this; and please forgive me for stressing what may seem an obvious point. G can be viewed also as $\langle x \rangle \times H$, and corresponding to this we may view R in two ways. First we may write it as

$$R = S\langle x \rangle .$$

Viewing it this way one may hope to prove some results because of the simplicity of the group $\langle x \rangle$. Second, we may view it as $R = J\langle x \rangle H$, or

$$R = J_1 H .$$

On the face of it, it does not look as though investigating R as J_1H could be of much help, or indeed be any easier than investigating JH itself. The point is that the answers one can get about simple S-modules mean, when they are applied to J_1H rather than JH, that one has information about the cycle $\langle x \rangle$ which can be used, when we view R in the first way to deduce information about S. It is looking at R in both ways which is the key to the situation. The analogy with polynomial rings is clear: $J[X_1, X_2, \ldots, X_n]$ can be viewed either as $J[X_1][X_2, \ldots, X_n]$, where inductions on the number of variables is in order, or as $J[X_1, \ldots, X_{n-1}][X_n]$ where only one variable is involved.

4. The proof depends upon two results which I hardly have time to prove, so that I will state them and discuss them very briefly. The first is

(A) Suppose $L \triangleleft_{+} S$ and λ is in S. Suppose that $\lambda L \leq L$ and that every simple R-module is finitely generated as an S-module. (*)

(i) If, for all maximal right ideals T of S containing L, there exists r with $\lambda^r \in T$, with $r = r(T)$, then $R = LR + (x-\lambda)R$.

(ii) If $R = LR + (x-\lambda)R$ then $\lambda^r \in L$ for some r.

The proof corresponds exactly with the way the Strong Nullstellensatz is proved in many books on commutative algebra. The point being that if we write M for S/L we have a one-generator S-module and the condition $\lambda L \leq L$ means that $L + \sigma \rightarrow L + \lambda\sigma$ is a module endomorphism. It commutes with the action of S. This allows us to blow S/L up into R/LR and then make x, which commutes with H, act like λ by factoring with $(x-\lambda)R$. There is a more general version where one starts with any finitely generated S-module M and λ is an element of $End_S(M)$; R/LR is then replaced by $M \otimes_S R$, but I wish to restrict our attention to the one-generator case. Now (A) is very relevant to our ideal version of the Strong Nullstellensatz. If we take S^ϕ to be S/L with $L \triangleleft S$, then $\lambda L \leq L$ is automatic. If $L + \lambda$ lies in Jac(S^ϕ) then λ lies in every T and consequently we deduce that some power of λ is in L. In other words Jac(S^ϕ) is a nil ideal of S^ϕ. But S^ϕ, like S, has Max-r and nil ideals in such rings are nilpotent. Hence the Strong version follows directly from (A), provided that (*) holds. Now G, like H, is certainly polycyclic by finite, and (*) would surely hold if Philip Hall's problem were actually true. However, the solution to that problem is quite irrelevant for the purposes of establishing (*). To establish (*) we turn to the second way of viewing R, viz as J_1H, which is just the group ring of H over a slightly bigger absolutely capital Jacobson ring, by Krull's version. The result is

(B) Suppose J is Jacobson and H is polycyclic by finite. If M is any simple JH-module then there is a maximal ideal U of J with $MU = 0$.

In other words M is naturally a vector space over the capital J/U of J. Of course, one has to have such a result even to make sense of the weak version of the theorem. Now (B) is the result which Philip Hall proved in the 1959 paper about polycyclic groups, except that he considered only the cases $J = \mathbb{Z}$ and $J = K\langle x\rangle$. His techniques relied upon another property in common between $\mathbb{Z}$ and $K\langle x\rangle$: they are both principal ideal domains. However the reliance upon that property is not essential. To cut rather a long story short, one might say that to prove (B) you go carefully through Hall's work, using much the same sort of arguments but eschewing principal ideal domains. Anyhow, what is relevant of this result for to-day is the corollary.

(C) If J is an absolutely capital Jacobson ring and M a simple R-module then $M(x^n - 1) = 0$ for some $n > 0$. (Consequently (*) holds).

For, view R as J_1H and appeal to (B). There is some maximal ideal U_1 of J_1 with $MU_1 = 0$. Since J_1 is absolutely capital by Krull's version, J_1/U_1 is absolute and x being invertible must satisfy $x^n \equiv 1 \pmod{U}$, as required. Then $M = \mu R = \mu \sum_r x^r S$, viewing R as $S\langle x\rangle$. Hence $M = \mu S + \mu xS + \dots + \mu x^{n-1}S$.

5. So this establishes the strong part of the theorem. The weak part is more difficult. We must consider a capital S/Y of S, so that Y is a maximal ideal of S. We must show that S/Y is finite. Let

$D = (1 + Y) \cap H$ be the kernel of $h \to Y + h$, the homomorphism of H into the unit group of S/Y. Obviously S/Y is a capital of H/D, so we may suppose that D is trivial. With this assumption we shall actually show that <u>H is finite</u>. This will be enough, for if we take a maximal right ideal T of S containing Y, then $M = S/T$ is a simple JH-module which, from (B), is a vector space over some capital of J. Consequently, so is S/Y; but it is obviously finitely generated, since H is finite.

At this point we need to know that H will be finite if every Abelian normal subgroup is finite. So let $1 = A' \triangleleft A \triangleleft H$. We must show that A is finite. Since H has Max-s, all that is necessary is to show that A is periodic. Suppose therefore that a is in A. The idea now is this: Y is an ideal of S and therefore $Y^h = h^{-1}Yh$ for all h in Y. We go above Y to a right ideal X of S maximal with respect to satisfying $X^A = \sum_{\alpha \in A} X^\alpha = X$, and then above X to a right ideal L of S maximal with respect to $L^a = L$, and we show that there is some positive n with $a^n - 1$ in L. This will be sufficient, because it will follow that $a^n - 1$ will lie in $\bigcap_{\alpha \in A} L^\alpha$. This is an A-invariant right ideal containing X and must therefore coincide with X. In other words $a^n - 1$ will be in X. If we set $B = (1 + X) \cap A$, then a^n will be in B. Consequently A/B will be periodic and hence finite. Therefore B will contain some power $C = A^m$ of A. This C has the advantage over B that it is normal in H and hence the right ideal $\bar{C}$ on the augmentation of C is a two-sided ideal. Now $\bar{C} \leq \bar{B} \leq X$. Hence $Y + \bar{C} \leq X$. By the maximality of Y it follows that $\bar{C} \leq Y$, so that C is contained in D which was assumed to be 1.

So all is done save for proving that $a^n - 1$ is in L. Let me postpone this momentarily and point out why all this is possible with maximal ideals Y and could not be succesful with maximal right ideals. It is because we know that S/Y, as an S-module, has left multiplication by elements of H as endomorphisms. Indeed, H appears as a subgroup of $End_S(S/Y)$ and we have built up to X which is a maximal A-invariant submodule and then to L which is a maximal a invariant submodule. At each stage we have a lot of group elements which appear as endomorphisms. In the module rather than the capital case one has very little information about the endomorphisms in connexion with the group, except when the group is nilpotent when the central elements acting on the right are, of course, module endomorphisms. This is why the module version of the 1961 result really works. What information we have about the endomorphisms ring will come out of finishing the proof of this theorem. Let me state it as

(D) Suppose λ is in S and that L is a right ideal of S. If L is maximal with respect to the conditions $\lambda L \leq L$ and $\lambda^r \notin L$ for any r, then there exists $n > 0$ with $\lambda^n - 1$ in L.

Our result will follow by setting $\lambda = a^{-1}$ and noticing that $L^a = L$ is equivalent with $a^{-1}L \leq L$, since S has Max-r.

To prove (D) we go back to $G = H \times \langle x \rangle$ and again consider (A). By (ii), $R > LR + (x - \lambda)R$, so there is a maximal right ideal V of R above $LR + (x - \lambda)R$. It follows from the maximality conditions on L that $L = V \cap S$. By the result (C), $(R/V)(x^n - 1) = 0$. Since $x - \lambda$ is in V it follows that $\lambda^n - 1$ is in L, as required.

This completes the proof of the theorem, but one may from (D) deduce

(E) If M is a simple S-module then its centralizer is an absolute field.

For if L is a maximal right ideal of S with $M \cong S/L$, then the centralizer of M may be thought of as $n(L)/L$, where $n(L)$ is the set of all those λ such that $\lambda L \leq L$. It is a division ring without more ado, so that if λ is not in L, then both conditions of (D) are satisfied and hence $\lambda^n - 1$ is in L for some $n > 0$. In other words $n(L)/L$ has all its elements roots of unity and it is therefore an absolute field.

Perhaps I could point out that similar methods will establish

(F) If M is a finitely generated S-module and λ is in $\mathrm{End}_S(M)$ then either $M\lambda^n = 0$ for some n or else there exists $n > 0$ with $M(\lambda^n - 1) < M$,

and also

(G) If M is a simple S-module with centralizer K and if M is finite dimensional over K then it is finite dimensional over a capital of J.

This is another way of putting precisely, what one knows at once about this problem of Hall: that what one has to do is to prove that either the centralizer of M is big or else the group H is small.

<u>Note.</u> What preceeds this note is more or less exactly what I said in the first two third's of my talk. The last third concerned the solution of Philip Hall's problem in certain special cases. This discussion led to a number of open questions with which I finished. I have suppressed this section. Shortly after the conference, thanks to some timely help from D.S. Passman, I managed to solve Hall's problem completely. In the language of the talk the result is

(F) Any simple S-module is finite dimensional over a capital of J.

This supercedes a certain amount of what I have said above, but it is considerably more difficult than the corresponding theorem about the capitals of S. It certainly disposes of most of the questions I asked at the end of my talk, hence the suppression. A full account will appear in the Proceedings of the London Mathematical Society in due course.

SQ-Universal 1-Relator Groups

G. S. Sacerdote

A countable group G is SQ-universal if for each countable group H, there is a quotient Q of G in which H can be embedded.

A group is a 1-relator group if it has a presentation by generators and relations involving only one relation.

Theorem: Let G be a countable 1-relator group. Then either G is SQ-universal or else G is one of the following: (i) cyclic

(ii) metabelian and isomorphic to $\langle a,b;\ a^{-1}b^{m}a=b^{n}\rangle$ where either $|m|=1$ or $|n|=1$.

Corollary: The Baumslag-Solitar group $\langle a,b;a^{-1}b^{2}a=b^{3}\rangle$ is SQ-universal.

A group is deficient if it has a presentation in which the number of generators exceeds the number of relations by at least two. Clearly a deficient group with zero relations is SQ-universal, for it is free. By the preceding theorem a deficient group with one relation is SQ-universal.

Conjecture (Peter Neumann): Every finitely generated deficient group is SQ-universal.

The Cohomology of Pregroups

John R. Stallings

Introduction

If a group G can be expressed as a free product with amalgamated subgroup $A *_C B$, then a projective resolution of Z over ZG can be constructed simply out of resolutions for A, B and C [7]. The geometric analogue of this is that a K(G, 1) space can be concocted out of the union of a K(A, 1) and a K(B, 1) intersecting in a K(C, 1) [5], the proof using a Meyer-Vietoris sequence in the universal covering spaces [8].

The present paper extends this type of construction to the generalization of amalgamated free product which we call "the universal group of a pregroup" [6]. Here, U(P) is analogous to G and P to the amalgam $A \vee_C B$. What we do is to construct a K(U(P), 1) complex only using P; then we construct a universal covering complex W(U(P), 1) on which U(P) acts freely. These are complete semi-simplicial complexes which do not usually satisfy a Kan condition; the contractibility of W is therefore algebraic or topological, but not simplicial, and the proof of contractibility involves essential properties of pregroups.

1. Pregroups

We recall here the basic facts about pregroups as described more completely in [6].

A pregroup consists of a set P, containing a distinguished element denoted 1, each element $p \in P$ has a unique inverse $p^{-1} \in P$, and to each pair of elements $p, q \in P$ there is defined at most one product $pq \in P$, such that

(i) $1p = p1 = p$, always defined.

(ii) $pp^{-1} = p^{-1}p = 1$, always defined.

(iii) If pq is defined, then $q^{-1}p^{-1}$ is defined and equal to $(pq)^{-1}$.

(iv) Supposing ab and bc are defined, then $a(bc)$ is defined if and only if $(ab)c$ is defined, in which case the two are equal.

(v) If ab, bc and cd are defined, then either $a(bc)$ or $(bc)d$ is defined.

It can be seen that a pregroup is one of the diverse types of structure studied by Baer [1].

Now, it is a theorem [6] that each pregroup P can be embedded in a universal group $U(P)$, which is in a sense the largest group that P can generate. If $g \in U(P)$, then g can be written as a product

$$g = p_1 p_2 \cdots p_n, \quad \text{for } p_1, p_2, \ldots, p_n \in P$$

such that no $p_i p_{i+1}$ is defined in P itself. Such an n-tuple $(p_1, \ldots, p_n)$ is said to be a reduced word representing g.

A basic theorem [6] is: Two reduced words $(p_1, \ldots, p_n)$, $(q_1, \ldots, q_m)$ represent the same element $g \in U(P)$ if and only if:

$$m = n, \text{ and}$$

There exist $a_1, \ldots, a_{n-1} \in P$, such that if we say $a_0 = a_n = 1$, then $p_i a_i$ and $a_{i-1}^{-1} p_i$ are defined in P, and $q_i = a_{i-1}^{-1} p_i a_i$, for all i.

Another way of thinking of this is to define an equivalence relation on reduced words, two words being equivalent if and only if they have the same length and one is obtainable from the other by the above sort of interleaving. Then $U(P)$ is, as a set, in one-to-one correspondence with the set of equivalence classes of reduced words; the product in $U(P)$ corresponds concatenation of reduced words followed by a reduction process if the concatenation is not reduced.

In particular, the number n, uniquely associated to g by the possibility of representing g by a reduced word of length n, will be called the length of g and denoted by $\ell(g)$. The elements of P are characterized by $\ell(p) = 1$. Note that in particular $\ell(1) = 1$.

The rest of this section is devoted to the details which will be used in the proof of our results.

1.1. <u>Definition:</u> For each $g \in G$, we choose a reduced word representing g, say $(p_1, \ldots, p_n)$, such that $g = p_1 p_2 \ldots p_n$. We then define $\lambda(g) = p_1 p_2 \ldots p_{n-1}$, <u>the left segment of</u> g. An exceptional case is when $\ell(g) = 1$, i.e. $g \in P$, in which case we require $\lambda(g) = 1 \in P$. Of course, there is no canonical way of defining $\lambda(g)$; hence this definition relies on the axiom of choice.

1.2. <u>Proposition:</u> If $h, g \in G$ and $h^{-1}g \in P$ and $\ell(h) \leq \ell(g)$, then

(i) $\ell(h) \geq \ell(g) - 1$, and

(ii) $g^{-1}\lambda(g) \in P$, and

(iii) $h^{-1}\lambda(g) \in P$.

<u>Proof:</u>

(i) Represent h by a reduced word,

$$h = q_1 q_2 \ldots q_m .$$

Then $g = h \cdot (h^{-1}g) = q_1 q_2 \ldots q_m \cdot (h^{-1}g)$. This shows that g can be represented by a word, not necessarily reduced, of length $\ell(h) + 1$. After reducing, g will be represented by a word of length $\leq \ell(h) + 1$, or $\ell(g) \leq \ell(h) + 1$.

(ii) $g^{-1}\lambda(g) = p_n^{-1} \in P$.

(iii) $h^{-1}\lambda(g) = (h^{-1}g) \cdot (g^{-1}\lambda(g))$, where the two factors here belong to P, so that if $h^{-1}\lambda(g)$ were not to belong to P, the product of $(h^{-1}g) \cdot (g^{-1}\lambda(g))$ would not be in P. Then the inverse product,

$$(\lambda(g)^{-1}g) \cdot (g^{-1}h)$$

would be the product of two elements of P whose product would not be in P. Represent g by $p_1 p_2 \ldots p_{n-1} p_n$ where $p_n = \lambda(g)^{-1}g$. Then

$$h = \lambda(g) \cdot (\lambda(g)^{-1}g) \cdot (g^{-1}h)$$

$$= p_1 p_2 \ldots p_{n-1} \cdot (\lambda(g)^{-1}g) \cdot (g^{-1}h)$$

would be represented by $(p_1, p_2, \ldots, p_{n-1}, \lambda(g)^{-1}g, g^{-1}h)$ which would be a reduced word of length $n+1$. Then $\ell(h) = n+1 > \ell(g) = n$, and this would be a contradiction with the hypothesis $\ell(h) \leq \ell(g)$. Therefore in fact $h^{-1}\lambda(g) \in P$.

2. Construction of K

Let P be a fixed pregroup.

By an n-dimensional simplex of K is meant an n-tuple $\sigma = (p_1, \ldots, p_n)$ of elements of P such that for all i, j, we have $p_i^{-1}p_j$ defined in P.

We shall define the faces and degeneracies of this simplex:

The ith face, $\partial_i\sigma$:

$\partial_0\sigma = (p_1^{-1}p_2, \ldots, p_1^{-1}p_n)$. That is, omit the first term and multiply all the other terms by the inverse of the first term.

$\partial_i\sigma = (p_1, \ldots, \hat{p}_i, \ldots, p_n)$, for $i > 0$. That is, omit the ith term.

The ith degeneracy, $d_i\sigma$:

$d_0\sigma = (1, p_1, \ldots, p_n)$, place 1 in front of σ

$d_i\sigma = (p_1, \ldots, p_{i-1}, p_i, p_i, p_{i+1}, \ldots, p_n)$, for $i > 0$.

That is, duplicate the ith term.

It is easily checked that if σ is a simplex, so is $\partial_i\sigma$ and $d_i\sigma$, and that these operators satisfy the relations for a complete semi-simplicial complex [3]. This c. s. s. will be denoted K(P).

In the above, the unique 0-dimensional simplex is the unique, empty, 0-tuple.

We note an alternative construction of K(P).

Let the n-tuple with brackets, $[q_1, q_2, \ldots, q_n]$ correspond to the n-tuple with parentheses $(q_1, q_1q_2, \ldots, q_1q_2 \cdots q_n)$, where the ith term in the latter n-tuple is to be the product of the first i terms in the former. The bracket form corresponding to $(p_1, \ldots, p_n)$ is then $[p_1, p_1^{-1}p_2, p_2^{-1}p_3, \ldots, p_{n-1}^{-1}p_n]$. We interpret the terms as belonging to U(P).

Then it is easy to see that, for $[q_1, \ldots, q_n]$ to correspond to simplex, it is necessary and sufficient that for all $i \leq j$, the segmental product $q_i q_{i+1} \dots q_j$ belongs to P.

The corresponding face and degeneracy formulas are:

$$\partial_0[q_1, \dots, q_n] = [q_2, \dots, q_n]$$
$$\partial_i[q_1, \dots, q_n] = [q_1, \dots, q_{i-1}, q_i q_{i+1}, q_{i+2}, \dots, q_n], \text{ for } 1 \leq i < n$$
$$\partial_n[q_1, \dots, q_n] = [q_1, \dots, q_{n-1}]$$
$$d_0[q_1, \dots, q_n] = [1, q_1, \dots, q_n]$$
$$d_i[q_1, \dots, q_n] = [q_1, \dots, q_i, 1, q_{i+1}, \dots, q_n] \text{ for } i > 0.$$

It should be remarked that $K(P)$ is defined exactly as the standard $K(\Pi, 1)$ [2], and that our parenthesized and bracketed descriptions correspond exactly to two standard descriptions of $K(\Pi, 1)$. The only difference is that there is a test as to whether an n-tuple is a simplex in the case of $K(P)$; this is that $p_i^{-1} p_j$ is defined in P; this is what is necessary in order for the 0th face to be meaningful.

In case $P = A \cup_C B$ is an amalgam of two groups along a subgroup, it is clear that a simplex of $K(P)$ is just a simplex either of $K(A, 1)$ or of $K(B, 1)$, so that $K(P)$ is exactly the union of $K(A, 1)$ and $K(B, 1)$. The common simplexes form $K(C, 1)$.

3. Construction of W

Let P be a fixed pregroup, and $U(P) = G$ its universal group.

By an n-dimensional simplex of W is meant an $(n+1)$-tuple $\Delta = (g_0/g_1/\dots/g_n)$ of elements of G, such that for all i and j, we have $g_i^{-1} g_j \in P$.

We define the ith face of Δ by deleting the ith term:

$$\partial_i \Delta = (g_0/\dots/\hat{g}_i/\dots/g_n)$$

We define the ith degeneracy of Δ by duplicating the ith term:

$$d_i \Delta = (g_0/\dots/g_{i-1}/g_i/g_i/g_{i+1}/\dots/g_n)$$

The totality of simplexes Δ with these operations obviously forms a complete semi-simplicial complex which we denote by $W(P)$.

3.1. The group G acts freely on $W(P)$ on the left, by the action:

$$h(g_0/\ldots/g_n) = (hg_0/\ldots/hg_n),$$

every term being multiplied by h. The quotient complex may be identified with $K(P)$; the quotient mapping $\varphi: W(P) \longrightarrow K(P)$ is given by

$$\varphi(g_0/g_1/\ldots/g_n) = (g_0^{-1}g_1, \ g_0^{-1}g_2, \ \ldots, \ g_0^{-1}g_n),$$

that is, omit the initial term and multiply all the others by its inverse.

Now we define, as usual, the chain complex, with Z coefficients, of $W(P)$:

C_n is the free Z-module on the set of n-dimensional simplexes of $W(P)$.

$\partial : C_n \longrightarrow C_{n-1}$ is the homomorphism, defined on the basis of C_n by the formula:

$$\partial(\Delta) = \sum_{i=0}^{n} (-1)^i \partial_i \Delta .$$

$\varepsilon : C_0 \longrightarrow Z$, the augmentation, is defined on the basis of C_0 by $\varepsilon(\Delta) = 1$.

The complex $\mathcal{C}(W(P))$ consists of $\{Z, C_0, C_1, \ldots\}$ plus all the homomorphisms $\partial : C_n \longrightarrow C_{n-1}$ and ε.

3.2. It easily follows from 3.1 that the chain complex of $W(P)$ forms a complex of free ZG-modules over Z, and, of course, the homology groups of this chain complex are the reduced homology groups of $W(P)$.

Our principal aim is to show this complex to have 0 homology, so that we have a resolution of free ZG-modules over Z. We shall make a brief digression on a bit-by-bit way of finding a contracting homotopy.

4. Slow contraction

Let (C, ∂) be a chain-complex. We omit dimensional indices as being irrelevant to this discussion; we think of C as an abelian group and $\partial : C \longrightarrow C$ an endomorphism, whose composition with itself, $\partial^2 = 0$. A chain endomorphism $f : C \longrightarrow C$ is an endomorphism of abelian groups such that $f\partial = \partial f$.

4.1. If $\tau : C \longrightarrow C$ is any abelian group endomorphism, then $\partial\tau + \tau\partial$ is a chain endomorphism.

For, $\partial(\partial\tau + \tau\partial) = (\partial\tau + \tau\partial)\partial = \partial\tau\partial$.

Suppose $\tau : C \longrightarrow C$ is as above, and $f = 1 - (\partial\tau + \tau\partial)$, where $1 : C \longrightarrow C$ is the identity chain-endomorphism. Then we say, as usual, that f is homotopic to 1 with homotopy τ.

4.2. Lemma. Suppose that f is homotopic to 1 with homotopy τ, and that C is generated as a group by a set $S \subset C$, such that for all $s \in S$ there exists $n = n(s)$, such that $f^n(s) = 0$. Then 0 is homotopic to 1 via a homotopy T, where, symbolically written:

$$T = \tau\left(\sum_{n=0}^{\infty} f^n\right).$$

Proof: Each element x of C is some finite linear combination of S. Since each n-fold composition f^n is an endomorphism, if we let N = maximum of $n(s)$ for s occurring in the expression for x, then it follows that $f^N(x) = 0$. We then define

$$T(x) = \tau(x + f(x) + f^2(x) + \ldots + f^{N-1}(x)).$$

This definition does not depend on N so long as N is sufficiently large.

Now the proof is just a computation:

$$\begin{aligned}
&(\partial T + T\partial)(x) \\
&= \partial\tau(x + f(x) + \ldots + f^{N-1}(x)) + \tau(\partial x + f(\partial x) + \ldots + f^{N-1}(\partial x)) \\
&= \partial\tau(x + f(x) + \ldots + f^{N-1}(x)) + \tau\partial(x + f(x) + \ldots + f^{N-1}(x)) \\
&= (\partial\tau + \tau\partial)(x + f(x) + \ldots + f^{N-1}(x)) \\
&= (1 - f)(1 + f + \ldots + f^{N-1})(x) \\
&= (1 - f^N)(x) \\
&= x = (1 - 0)(x).
\end{aligned}$$

Here we used the facts: f^k is a chain-endomorphism for all k, and $\partial\tau + \tau\partial = 1 - f$.

This is, in a sense, a lemma in the tradition of the Riemann mapping theorem, Eilenberg's projective module swindle, and Mazur's sphere-imbedding theorem, in that all difficulties are pushed to ∞ where they vanish.

5. Algebraic contractibility of W(P)

Let $\mathcal{C} = (\{C_n\}, Z, \{\partial\}, \varepsilon)$ be the chain-complex of W(P).

Let $\Delta = (g_0/g_1/\ldots/g_n)$ be an n-simplex; therefore each $g_i^{-1}g_j \in P$ and $g_i \in U(P)$. Then, by (1.2)(i), either all the lengths of terms of Δ are equal: $\ell(g_i) = k$ for all i; or the lengths comprise two adjacent integers: $\ell(g_i) = k$ or $k + 1$ for all i.

In the former case, when all $\ell(g_i)$ are equal, to k, say, define $h_\Delta = \lambda(g_0)$, the left segment of the initial term in Δ.

In the latter case, when there are two lengths, k and $k + 1$, let i be the least subscript such that $\ell(g_i) = k + 1$, and define $h_\Delta = \lambda(g_i)$, the left segment of the first term in Δ having the longer length.

It follows from (1.2)(iii) that if $\Delta = (g_0/g_1/\ldots/g_n)$, then $h_\Delta^{-1}g_i \in P$ for all i, and hence we define

$$\tau(\Delta) = (h_\Delta/g_0/g_1/\ldots/g_n)\,,$$

which is a simplex of W(P) of dimension $n + 1$. This is obtained by adding in the initial position, the left segment of the first term of Δ having longest length.

We now extend τ linearly, so that $\tau : C_n \longrightarrow C_{n+1}$ is a homomorphism, and define $\tau : Z \longrightarrow C_0$ by: $\tau(1) = (1)$.

It will be shown that τ is a slow contraction, in the sense of the previous section.

5.1. Let $\Delta = (g_0/g_1/\ldots/g_n)$ and let g_i be the first term with the longest length. Then

$$\tau\partial_j\Delta = \partial_{j+1}\tau\Delta \quad \text{for } j \neq i.$$

__Proof:__ $\partial_j\Delta = (g_0/g_1/\ldots/\hat{g}_j/\ldots/g_n)$. Thus, the first term in $\partial_j\Delta$ with longest length is still g_i, if $i \neq j$. Thus $\tau\partial_j\Delta = (h_\Delta/g_0/\ldots/\hat{g}_j/\ldots/g_n) = \partial_{j+1}\tau\Delta$.

5.2. Let Δ be as above. Then

$$\begin{aligned}
(\partial\tau + \tau\partial)\Delta &= \sum_{j=0}^{n+1} (-1)^j \partial_j \tau\Delta + \tau\Big(\sum_{j=0}^{n} (-1)^j \partial_j \Delta\Big) \\
&= \Delta + \sum_{j=0}^{n} \{(-1)^{j+1}\partial_{j+1}\tau\Delta + (-1)^j \tau\partial_j\Delta\} \\
&= \Delta + (-1)^{i+1}(\partial_{i+1}\tau\Delta - \tau\partial_i\Delta),
\end{aligned}$$

since $\partial_0\tau\Delta = \Delta$, and all but one term in the summation is zero by 5.1.

We now define $f : \mathcal{C} \longrightarrow \mathcal{C}$, which will be a chain endomorphism, by 4.1, by:

$$f = 1 - (\partial\tau + \tau\partial).$$

Then according to the above:

$$(*) \qquad \begin{aligned}
f(\Delta) &= (-1)^i(\partial_{i+1}\tau\Delta - \tau\partial_i\Delta) \\
&= (-1)^i\{(h_\Delta/g_0/\ldots/\hat{g}_i/\ldots/g_n) \\
&\qquad - (h_{\partial_i\Delta}/g_0/\ldots/\hat{g}_i/\ldots/g_n)\}.
\end{aligned}$$

(5.3). If the lengths of all terms of Δ are 1, then $f(\Delta) = 0$. This happens because $h_\Delta = h_{\partial_i\Delta} = 1$ in this case, because $\lambda(p) = 1$ for $p \in P$, by definition.

(5.4). On Z, i.e., on the (-1)-dimensional part of C, f is zero. Here,

$$\begin{aligned}
f(1) &= 1 - (\varepsilon\tau + \tau\cdot 0)(1) \\
&= 1 - \varepsilon((1)) = 0 ;
\end{aligned}$$

(5.5). Define the complexity of Δ thus: $c((g_0/\ldots/g_n)) = \ell(g_0) + \ldots + \ell(g_n) - (n+1)$.

If $c(\Delta) > 0$, then $f(\Delta)$ is the difference of two simplexes of complexity $< c(\Delta)$.

Proof: We note that for $c(\Delta) > 0$ we must have $\ell(g_i) > 1$ for g_i the first term of longest length in Δ. Thus $\ell(h_\Delta) = \ell(\lambda(g_i)) = \ell(g_i) - 1$. Clearly $\ell(h_{\partial_i \Delta}) \leq \ell(h_\Delta)$, and therefore the complexity of each simplex of (*) is obtained from $c(\Delta)$ by subtracting $\ell(g_i)$ and adding $\ell(h_\Delta)$ or $\ell(h_{\partial_i \Delta})$, both of which are $< \ell(g_i)$. Thus $c(\partial_{i+1} \tau \Delta) = c(\Delta) - 1$, and

$$c(\tau \partial_i \Delta) \leq c(\Delta) - 1 .$$

(5.6). Either $f^k(\Delta) = 0$, or $f^k(\Delta)$ is a linear combination of simplexes of complexity $\leq c(\Delta) - k$.

This is provable by induction on k, using 5.5 and 5.3.

(5.7). For some k, $f^k(\Delta) = 0$.

By (5.6) and (5.3) it is sufficient to take $k = c(\Delta) + 1$.

<u>Theorem:</u> The chain-complex of $W(P)$ is acyclic.

<u>Proof:</u> The existence of a contracting homotopy follows from 5.7 and 4.2. Q.E.D.

This signifies that the chain-complex of $W(P)$ is a free $Z(U(P))$-resolution of Z.

6. Topological contractibility of W

We now suppose that $W(P)$ has some nice geometric realization. In all probability, there is a topological contraction of $W(P)$ over itself to a point which exactly mimics the algebraic contraction established in the previous section. However there seem to be some unpleasant geometric details to this line of reasoning.

We shall suppose of $W(P)$, in geometric realization, that it has the homotopy type of a CW-complex, the homolpgy groups of $\mathcal{C}(W(P))$--hence, by section 5, the homology groups of a point, and that the fundamental group of $W(P)$ can be described as the edge-path group [4]. Then by a theorem of

Whitehead, the contractibility of $W(P)$ is implied by the triviality of its edge-path group.

We take the 0-simplex (1) as base point.

The edge-path group consists of equivalence classes of coherent based paths.

A coherent based path is an n-tuple, for $n \geq 0$, of the form $(E_1^{\varepsilon_1}, E_2^{\varepsilon_2}, \ldots, E_n^{\varepsilon_n})$, where $\varepsilon_i = \pm 1$, the symbols E_i are 1-simplexes; letting $\eta_i = \frac{1-\varepsilon_i}{2}$ and $\theta_i = \frac{1+\varepsilon_i}{2}$, we require

$$\partial_{\eta_i}(E_1) = (1)$$
$$\partial_{\theta_i}(E_n) = (1)$$
$$\partial_{\theta_i}(E_i) = \partial_{\eta_i}(E_{i+1})$$

That is to say, we regard E_i as a path going from $\partial_0 E_i$ to $\partial_1 E_i$, and E_i^{-1} as a path going in the reverse direction. Then a coherent based path is a concatenation of these basic units, starting and ending at (1) and arranged coherently.

The equivalence relation is generated by 2-simplexes as follows:

Let Δ be a 2-simplex. Then we say

$$(\partial_1\Delta) \sim_\Delta (\partial_0\Delta, \partial_2\Delta)$$
$$(\partial_2\Delta) \sim_\Delta (\partial_0\Delta^{-1}, \partial_1\Delta)$$
$$(\partial_0\Delta) \sim_\Delta (\partial_1\Delta, \partial_2\Delta^{-1})$$

as well as

$$(\partial_1\Delta^{-1}) \sim_\Delta (\partial_2\Delta^{-1}, \partial_0\Delta^{-1}) \text{ etc.}$$

We also say $d_0\rho \sim_\rho \Lambda$, the empty path, if ρ is any 0-simplex and $d_0\rho$ is its degeneracy.

Then let $\xi F \mathcal{V}$ be any coherent based path, where ξ is a coherent path starting at (1), F is one of the above small paths on the boundary of a

2-simplex, where $F \sim_\Delta F'$, and $\mathcal{G}$ is a coherent path ending at (1). We say $\mathcal{E}F\mathcal{G} \approx_\Delta \mathcal{E}F'\mathcal{G}$.

The equivalence relation on the coherent based paths is the equivalence relation generated by all $\approx_\Delta$ and $\approx_P$. The equivalence classes form a group whose multiplication is defined by concatenation. This is the edge-path group, which in the present instance we wish to show is trivial

6.1. Let $(E_1^{\varepsilon_1}, \ldots, E_n^{\varepsilon_n})$ be a coherent based path in $W(P)$. It is equivalent to a path in which all the exponents are positive.

For, let $(g_0/g_1) = E_i$ be a 1-simplex. We have $g_0, g_1 \in U(P)$ and $g_1^{-1}g_0 \in P$. Let

$$\Delta = (g_0/g_1/g_0) .$$

Then Δ is a 2-simplex, and $\partial_0\Delta = (g_1/g_0)$,

$$\partial_1\Delta = d_0(g_0), \; \partial_2\Delta = E_i .$$

We have $(E_i^{-1}) = (\partial_2\Delta^{-1}) \sim (\partial_1\Delta^{-1}, \partial_0\Delta) = (d_0(g_0)^{-1}, (g_1/g_0)) \sim (g_1/g_0)$.

In other words, an occurrence of $(g_0/g_1)^{-1}$ can be replaced by (g_1/g_0).

6.2. Now let $(E_1, \ldots, E_n)$ be a coherent based path with positive exponents. The condition $\partial_0 E_1 = (1)$ means $E_1 = (p_1/1)$ where $p_1 \in P$. Similarly $\partial_1 E_n = (1)$ means $E_n = (1/g_{n-1})$ where $g_{n-1} \in P$.

The condition $\partial_1 E_i = \partial_0(E_{i+1})$ means $E_i = (g_{i-1}p_i/g_{i-1})$ and $E_{i+1} = (g_i p_{i+1}/g_i)$, where $p_i, p_{i+1} \in P$ and $g_i = g_{i-1}p_i$.

Thus, a coherent based path is completely determined by a sequence $\{p_1, \ldots, p_n\}$ of terms in P, so that

$$E_1 = (p_1/1)$$

$$E_i = (p_1p_2 \cdots p_i/p_1p_2 \cdots p_{i-1})$$

and $p_1p_2 \cdots p_n = 1$, since

$$E_n = (1/p_1p_2 \cdots p_{n-1}).$$

If the length n of the sequence is 1, then $p_1 = 1$, the path consists only of E_1 which is degenerate and so is equivalent to the empty path.

If $n > 1$, then $(p_1, \ldots, p_n)$ cannot be a reduced word, since 1 is not represented by any reduced word of length > 1. Thus, for some i, $p_i p_{i+1} \in P$. Let $g = p_1 p_2 \ldots p_{i-1}$. Then $E_i = (gp_i/g)$, $E_{i+1} = (gp_i p_{i+1}/gp_i)$. Let $\Delta = (gp_i p_{i+1}/gp_i/g)$; this is a simplex because $p_i p_{i+1} \in P$. We have now:

$$(E_i, E_{i+1}) = (\partial_0\Delta, \partial_2\Delta) \sim (\partial_1\Delta) = (gp_i p_{i+1}/g).$$

Thus the original coherent based path is equivalent to one of length $n - 1$; the new path reflects a reduction of the word $(p_1, \ldots, p_n)$.

Thus, by induction on its length n, any coherent based path in $W(P)$ is equivalent to the empty path, and so the edge path group of $W(P)$ is trivial, and thus $W(P)$ is topologically contractible.

References

1. R. Baer, Free sums of groups and their generalizations. III., Amer. J. Math. 72 (1950), 647-670.

2. S. Eilenberg and S. MacLane, Cohomology theory in abstract groups. I, Ann. of Math., 48 (1947), 51-78.

3. D. M. Kan, A combinatorial definition of homotopy groups, Ann. of Math. 67 (1958), 282-312.

4. H. Seifert and W. Threlfall, Lehrbuch der Topologie, Leipzig (1934).

5. J. Stallings, A finitely presented group whose 3-dimensional integral homology is not finitely generated, Amer. J. Math. 85 (1963), 541-543.

6. J. Stallings, Group theory and three-dimensional manifolds, Yale University Press, New Haven (1971).

7. R. G. Swan, Groups of cohomological dimension one, J. Algebra 12 (1969), 585-601.

8. J. H. C. Whitehead, On the asphericity of regions in the 3-sphere, Fund. Math. 32 (1939), 149-166.

A Survey of SQ-Universality

Paul E. Schupp

A countable group G is said to be SQ-universal if for every countable group C there exists a quotient group, G_C, of G in which C can be embedded. (In short, every countable group is isomorphic to a subgroup of a quotient group of G.) The only totally obvious example of an SQ-universal group is the free group of rank $\aleph_0$. The theorem of Higman, Neumann, and Neumann [3] stating that every countable group can be embedded in a two generator group shows that the free group of rank two, and thus all non-abelian free groups, are SQ-universal.

Suppose that G is SQ-universal. Since some quotient of G embeds the free group of rank $\aleph_0$, G must contain free subgroups of rank $\aleph_0$. A theorem of Bernard Neumann [5] shows that there are $2^{\aleph_0}$ non-isomorphic two generator groups. Since any single quotient group of G can embed only countably many different two generator groups, G must have $2^{\aleph_0}$ non-isomorphic quotient groups. The property of being SQ-universal may, in a very rough sense, be considered as an indication of "largeness" or "freeness". P. Hall and P. Neumann [6] have independently shown.

Theorem 1. Suppose H is a subgroup of G

with $[G:H] < \infty$. Then G is SQ-universal if and only if H is SQ-universal.

Using the above, P. Neumann [6] shows that

Theorem 2. If G is a finitely generated Fuchsian group which does not have an abelian subgroup of finite index, then G is SQ-universal.

The most powerful method for showing groups to be SQ-universal is "small cancellation theory". Without entering into a discussion of these methods (See [10] for a general survey), we state some of the results on SQ-universality. The next theorem was proven by J. L. Britton in the fifties (unpublished) and was later independently proven (in essence) by F. Levin [4].

Theorem 3. If $G = H * K$ is any non-trivial free product except $\mathbb{Z}_2 * \mathbb{Z}_2$, then G is SQ-universal

It turns out that many free products with amalgamation are also SQ-universal. Let A be a subgroup of the group H. A pair $\{x_1, x_2\}$ of elements of H is said to be a blocking pair for A in H if x_1, x_2 are distinct, neither is in A, and the following two conditions are satisfied:

(i) $x_i^{\epsilon} x_j^{\delta} \notin A$, $1 \leq i,j \leq 2$, $\epsilon = \pm 1$, $\delta = \pm 1$, unless $x_0^{\epsilon} x_j^{\delta} = 1$,

(ii) if $a \in A$, $a \neq 1$, then $x_i^{\epsilon} a x_j^{\delta} \notin A$, $1 \leq i, j \leq \delta$, $\epsilon = \pm 1$, $\delta = \pm 1$.

The method for proving the existence of blocking pairs is usually the following. Suppose there is a subgroup B of H which has a non-trivial decomposition as a free product, $B = B_1 * B_2$, and $A \subseteq B_1$. Then any pair of distinct non-identity elements of B_2 is a blocking pair for A in H. Schupp [9] proves

Theorem 4. Let $G = (H * K; A = B)$ be a free product amalgamating proper subgroups A and B of H and K respectively. If there exists a blocking pair for A in H then G is SQ-universal.

M. Hall's work [1] shows that if H is a free group and A is a finitely generated subgroup of H, then there exists a subgroup B of finite index which contains A as a free factor. If A has infinite index in H, it follows that $B = A * B_2$ where B_2 is a non-trivial free group. Thus there exists a blocking pair for A in H. We thus have

Corollary 1. Let H be a free group, and let A be a finitely generated subgroup with infinite index. Let K be any group containing a proper subgroup B isomorphic to A. Then the free product with amalgamation $G = (H * K; A = B)$ is SQ-universal.

In [2], G. Higman used the group
$H = \langle a,b,c,d;\ b^{-1}ab = a^2,\ c^{-1}bc = b^2,\ d^{-1}cd = c^2,\ a^{-1}da = d^2 \rangle$
to establish the existence of finitely generated infinite simple groups. He showed that H is infinite but has no

non-trivial finite quotient groups. Thus H/M is infinite simple for M a maximal normal subgroup of H. It has been conjectured that H itself might be simple or close to it. One proves that H is infinite by showing that H has a decomposition as a free product with amalgamation. Surprisingly, it turns out that there is a blocking pair for the amalgamated subgroup and thus

<u>Corollary 2</u>. The Higman group H is SQ-universal.

A theorem analogous to Theorem 4 holds for HNN groups. Sacerdote and Schupp [8] show that

<u>Theorem 5</u>. Let $G = \langle H,t;\ t^{-1}At \underset{\varphi}{=} B \rangle$ be an HNN group. If there is an element $x \in G$ such that $x \notin A$, $x \notin B$, and $x^{-1}Bx \cap A = \{1\}$, then G is SQ-universal.

As a corollary of the theorem they deduce

<u>Corollary 3.</u> If $G = \langle a_1, \ldots, a_n;\ r \rangle$ is a group with a presentation having one defining relator and $n \geq 3$ generators, then G is SQ-universal.

Sacerdote [7] subsequently showed

<u>Theorem 6</u>. If G is a one relator group which is not metabelian, then G is SQ-universal.

Peter Neumann [6] has conjectured that if G is a group which has a presentation with r defining relations and at least $r + 2$ generators then G is SQ-universal. If $r = o$, the group is a non-abelian free group and the Higman-Neumann-Neumann theorem shows that G is SQ-universal.

Corollary 3 verifies the conjecture for $r = 1$. Few techniques for tackling the general case seem to be available at present. In preparing this talk I am indebted to the paper of Peter Neumann [6].

Bibliography

1.) Hall, M., Subgroups of finite index in free groups, Canadian J. Math., 1(1949), 187-190.

2.) Higman, G., A finitely generated infinite simple group, J. London Math Soc. 26 (1951), 61-64.

3.) Higman, G., Neumann B., and Hanna Neumann, Embedding theorems for groups, J. London Math. Soc. 24 (1949), 247-254.

4.) Levin, F., Factor groups of the modular group, J. London Math. Soc. 43 (1968), 195-203.

5.) Neumann, B. H., Some remarks on infinite groups, J. London Math. Soc. 12 (1937), 120-127.

6.) Neumann P., The SQ-universality of some finitely presented groups, to appear.

7.) Sacerdote, G., SQ-universality in one relator groups, J. London Math. Soc., to appear.

8.) Sacerdote, G., and Schupp, P., SQ-universality in HNN groups and one relator groups, J. London Math. Soc., to appear.

9.) Schupp, P., Small cancellation theory over free products with amalgamation, Math. Annalen 193 (1971), 255-264.

10.) Schupp, P., A survey of small cancellation theory, in Decision Problems and The Burnside Problem, North-Holland, 1972.

Vol. 212: B. Scarpellini, Proof Theory and Intuitionistic Systems. VII, 291 pages. 1971. DM 24,–

Vol. 213: H. Hogbe-Nlend, Théorie des Bornologies et Applications. V, 168 pages. 1971. DM 18,–

Vol. 214: M. Smorodinsky, Ergodic Theory, Entropy. V, 64 pages. 1971. DM 16,–

Vol. 215: P. Antonelli, D. Burghelea and P. J. Kahn, The Concordance-Homotopy Groups of Geometric Automorphism Groups. X, 140 pages. 1971. DM 16,–

Vol. 216: H. Maaß, Siegel's Modular Forms and Dirichlet Series. VII, 328 pages. 1971. DM 20,–

Vol. 217: T. J. Jech, Lectures in Set Theory with Particular Emphasis on the Method of Forcing. V, 137 pages. 1971. DM 16,–

Vol. 218: C. P. Schnorr, Zufälligkeit und Wahrscheinlichkeit. IV, 212 Seiten 1971. DM 20,–

Vol. 219: N. L. Alling and N. Greenleaf, Foundations of the Theory of Klein Surfaces. IX, 117 pages. 1971. DM 16,–

Vol. 220: W. A. Coppel, Disconjugacy. V, 148 pages. 1971. DM 16,–

Vol. 221: P. Gabriel und F. Ulmer, Lokal präsentierbare Kategorien. V, 200 Seiten. 1971. DM 18,–

Vol. 222: C. Meghea, Compactification des Espaces Harmoniques. III, 108 pages. 1971. DM 16,–

Vol. 223: U. Felgner, Models of ZF-Set Theory. VI, 173 pages. 1971. DM 16,–

Vol. 224: Revêtements Etales et Groupe Fondamental. (SGA 1). Dirigé par A. Grothendieck XXII, 447 pages. 1971. DM 30,–

Vol. 225: Théorie des Intersections et Théorème de Riemann-Roch. (SGA 6). Dirigé par P. Berthelot, A. Grothendieck et L. Illusie. XII, 700 pages. 1971. DM 40,–

Vol. 226: Seminar on Potential Theory, II. Edited by H. Bauer. IV, 170 pages. 1971. DM 18,–

Vol. 227: H. L. Montgomery, Topics in Multiplicative Number Theory. IX, 178 pages. 1971. DM 18,–

Vol. 228: Conference on Applications of Numerical Analysis. Edited by J. Ll. Morris. X, 358 pages. 1971. DM 26,–

Vol. 229: J. Väisälä, Lectures on n-Dimensional Quasiconformal Mappings. XIV, 144 pages. 1971. DM 16,–

Vol. 230: L. Waelbroeck, Topological Vector Spaces and Algebras. VII, 158 pages. 1971. DM 16,–

Vol. 231: H. Reiter, L^1-Algebras and Segal Algebras. XI, 113 pages. 1971. DM 16,–

Vol. 232: T. H. Ganelius, Tauberian Remainder Theorems. VI, 75 pages. 1971. DM 16,–

Vol. 233: C. P. Tsokos and W. J. Padgett. Random Integral Equations with Applications to Stochastic Systems. VII, 174 pages. 1971. DM 18,–

Vol. 234: A. Andreotti and W. Stoll. Analytic and Algebraic Dependence of Meromorphic Functions. III, 390 pages. 1971. DM 26,–

Vol. 235: Global Differentiable Dynamics. Edited by O. Hájek, A. J. Lohwater, and R. McCann. X, 140 pages. 1971. DM 16,–

Vol. 236: M. Barr, P. A. Grillet, and D. H. van Osdol. Exact Categories and Categories of Sheaves. VII, 239 pages. 1971, DM 20,–

Vol. 237: B. Stenström. Rings and Modules of Quotients. VII, 136 pages. 1971. DM 16,–

Vol. 238: Der kanonische Modul eines Cohen-Macaulay-Rings. Herausgegeben von Jürgen Herzog und Ernst Kunz. VI, 103 Seiten. 1971. DM 16,–

Vol. 239: L. Illusie, Complexe Cotangent et Déformations I. XV, 355 pages. 1971. DM 26,–

Vol. 240: A. Kerber, Representations of Permutation Groups I. VII, 192 pages. 1971. DM 18,–

Vol. 241: S. Kaneyuki, Homogeneous Bounded Domains and Siegel Domains. V, 89 pages. 1971. DM 16,–

Vol. 242: R. R. Coifman et G. Weiss, Analyse Harmonique Non-Commutative sur Certains Espaces. V, 160 pages. 1971. DM 16,–

Vol. 243: Japan-United States Seminar on Ordinary Differential and Functional Equations. Edited by M. Urabe. VIII, 332 pages. 1971. DM 26,–

Vol. 244: Séminaire Bourbaki – vol. 1970/71. Exposés 382–399. IV, 356 pages. 1971. DM 26,–

Vol. 245: D. E. Cohen, Groups of Cohomological Dimension One. V, 99 pages. 1972. DM 16,–

Vol. 246: Lectures on Rings and Modules. Tulane University and Operator Theory Year, 1970–1971. Volume I. X, 661 pages. DM 40,–

Vol. 247: Lectures on Operator Algebras. Tulane University Rin Operator Theory Year, 1970–1971. Volume II. XI, 786 pages. DM 40,–

Vol. 248: Lectures on the Applications of Sheaves to Ring Th Tulane University Ring and Operator Theory Year, 1970–1971 ume III. VIII, 315 pages. 1971. DM 26,–

Vol. 249: Symposium on Algebraic Topology. Edited by P. J. H VII, 111 pages. 1971. DM 16,–

Vol. 250: B. Jónsson, Topics in Universal Algebra. VI, 220 p 1972. DM 20,–

Vol. 251: The Theory of Arithmetic Functions. Edited by A. A. and D. L. Goldsmith VI, 287 pages. 1972. DM 24,–

Vol. 252: D. A. Stone, Stratified Polyhedra. IX, 193 pages. DM 18,–

Vol. 253: V. Komkov, Optimal Control Theory for the Dampi Vibrations of Simple Elastic Systems. V, 240 pages. 1972. DM

Vol. 254: C. U. Jensen, Les Foncteurs Dérivés de $\varprojlim$ et leur plications en Théorie des Modules. V, 103 pages. 1972. DM

Vol. 255: Conference in Mathematical Logic – London '70. Edit W. Hodges. VIII, 351 pages. 1972. DM 26,–

Vol. 256: C. A. Berenstein and M. A. Dostal, Analytically Un Spaces and their Applications to Convolution Equations. VII, pages. 1972. DM 16,–

Vol. 257: R. B. Holmes, A Course on Optimization and Bes proximation. VIII, 233 pages. 1972. DM 20,–

Vol. 258: Séminaire de Probabilités VI. Edited by P. A. Mey 253 pages. 1972. DM 22,–

Vol. 259: N. Moulis, Structures de Fredholm sur les Variété bertiennes. V, 123 pages. 1972. DM 16,–

Vol. 260: R. Godement and H. Jacquet, Zeta Functions of S Algebras. IX, 188 pages. 1972. DM 18,–

Vol. 261: A. Guichardet, Symmetric Hilbert Spaces and Relate pics. V, 197 pages. 1972. DM 18,–

Vol. 262: H. G. Zimmer, Computational Problems, Methods Results in Algebraic Number Theory. V, 103 pages. 1972. DM

Vol. 263: T. Parthasarathy, Selection Theorems and their Applica VII, 101 pages. 1972. DM 16,–

Vol. 264: W. Messing, The Crystals Associated to Barsott Groups: with Applications to Abelian Schemes. III, 190 pages. DM 18,–

Vol. 265: N. Saavedra Rivano, Catégories Tannakiennes. II, pages. 1972. DM 26,–

Vol. 266: Conference on Harmonic Analysis. Edited by D. G and R. L. Lipsman. VI, 323 pages. 1972. DM 24,–

Vol. 267: Numerische Lösung nichtlinearer partieller Differentia Integro-Differentialgleichungen. Herausgegeben von R. Ansorg W. Törnig, VI, 339 Seiten. 1972. DM 26,–

Vol. 268: C. G. Simader, On Dirichlet's Boundary Value Problem. pages. 1972. DM 20,–

Vol. 269: Théorie des Topos et Cohomologie Etale des Sch (SGA 4). Dirigé par M. Artin, A. Grothendieck et J. L. Verdie 525 pages. 1972. DM 50,–

Vol. 270: Théorie des Topos et Cohomologie Etle des Sch Tome 2. (SGA 4). Dirige par M. Artin, A. Grothendieck et J. L. V V, 418 pages. 1972. DM 50,–

Vol. 271: J. P. May, The Geometry of Iterated Loop Spaces. IX pages. 1972. DM 18,–

Vol. 272: K. R. Parthasarathy and K. Schmidt, Positive Definit nels, Continuous Tensor Products, and Central Limit Theore Probability Theory. VI, 107 pages. 1972. DM 16,–

Vol. 273: U. Seip, Kompakt erzeugte Vektorräume und Analys 119 Seiten. 1972. DM 16,–

Vol. 274: Toposes, Algebraic Geometry and Logic. Edited by. Lawvere. VI, 189 pages. 1972. DM 18,–

Vol. 275: Séminaire Pierre Lelong (Analyse) Année 1970–19 181 pages. 1972. DM 18,–

Vol. 276: A. Borel, Représentations de Groupes Localement pacts. V, 98 pages. 1972. DM 16,–

Vol. 277: Séminaire Banach. Edité par C. Houzel. VII, 229 1972. DM 20,–